I Fled Eternal Night
Escaped That Ghostly Lair
To Gain The Joy of Light.

George S. Campbell (2021),
Wither the Westerly Wind,
London: Austin Macauley.

Self-Different Fractals and Innovation

Academic Firm and the Entrepreneurial University in Epistemic Governance

David F. J. Campbell
University for Continuing Education Krems
Department for Higher Education Research
Center for Higher Education
Governance and Transformation
Krems, Austria
University of Applied Arts
Vienna, Austria

Elias G. Carayannis
George Washington University
School of Business
Washington, DC, United States of America

CRC Press
Taylor & Francis Group
Boca Raton London New York

CRC Press is an imprint of the
Taylor & Francis Group, an **informa** business
A SCIENCE PUBLISHERS BOOK

With Graphical Images by
Jürgen Pirecki-Giefing.

First edition published 2026
by CRC Press
2385 NW Executive Center Drive, Suite 320, Boca Raton FL 33431

and by CRC Press
4 Park Square, Milton Park, Abingdon, Oxon, OX14 4RN

CRC Press is an imprint of Taylor & Francis Group, LLC

Library of Congress Cataloging-in-Publication Data (applied for)

ISBN: 978-1-041-28191-7 (hbk)
ISBN: 978-1-041-13468-8 (pbk)
ISBN: 978-1-315-20923-4 (ebk)

DOI: 10.1201/9781315209234

Typeset in Times New Roman
by Prime Publishing Services

Dedication

This book we dedicate
To Our Families,
living on both sides of
The Atlantic Ocean.

Quote

Carayannis, Elias G./David F.J. Campbell (2020). The Knowledge and Innovation Principle of KITA PITA C2ITA, 1-5, in: Elias G. Carayannis (Editor-in-Chief): Encyclopedia of Creativity, Invention, Innovation and Entrepreneurship (https://doi.org/10.1007/978-1-4614-6616-1_200100-1)

Acknowledgment

We want to thank Vijay Primlani, Acquisitions Editor at Science Publishers, CRC PRESS, Boca Raton, Florida, United States. Through all the years he had and has supported us, encouraging us to continuously focus on the task of working on this book. *It was a long journey. It took us more than ten years to finally complete our book.*

We want to thank Jürgen Pirecki-Giefing for his masterly *Translation of the Fractal into Graphical Visualization. He made the fractals visible.*

Furthermore, we want to thank all our colleagues from our institutional networks, *with whom we engaged in so many fascinating intellectual discussions* over the many past years.

Their works are cited and reflected on in a dialog of thought and thoughts.

Of course, for possible errors in this manuscript, only we the authors alone are responsible.

Preface

In terms of a "dialectics of innovation", the "self-different fractals" are introduced as a philosophical principle. The interplay between "thesis" and "antithesis" can be disruptive, however, if managed or governed properly, such an interaction acts as a "synthesis" driver for creativity, co-creation, innovation, co-development and finally development. The need to govern innovation may mean to design the governance of innovation in a way so that it has the capability and capacity to also govern a "self-different fractal system of innovation".

Entrepreneurial university: The Mode 3 knowledge paradigms encourage universities to decide whether to focus on Mode 1, Mode 2, or a mix of both while finding innovative ways to integrate them. Academic Firm: The academic firm creates internally an "internal academic environment" or "internal academic ecosystem". The purpose of this is to simulate and emulate an "academic world within" a company, so to encourage and to facilitate similar motivations (as in the context of higher education) for engaging knowledge workers in (partially academic-style) research and innovation, where activities are connected with creativity-initiated and innovation-driven knowledge creation, knowledge production and knowledge application. Fortunately, this translates into commercially successful business opportunities. Epistemic Governance: epistemic governance refers explicitly to the underlying understandings (the underlying epistemic structure) of academia, for example of "quality", and specifically designs and implements policy and strategy, so as to create an interlinkage between governance and organizational development.

The future of higher education is reflected in the future of governance of and in higher education.

Contents

Acknowledgments *ix*
Preface *xi*

1. **Introduction: Self-Different Fractals and Innovation** **1**
2. **Governance of and in Higher Education** **8**
3. **The "Knowledge State": Governance of Higher Education in the Knowledge Society, Knowledge Economy and Knowledge Democracy** **19**
4. **Governance of Higher Education, "Epistemic Governance" and the Governance of Quality Enhancement in Higher Education Institutions** **26**
5. **Innovation: Linear and Non-Linear Models of Innovation and Innovation Networks of Entrepreneurial Universities and Academic Firms** **37**
6. **New Academic Careers** **47**
7. **The Academic Firm and Innovation** **51**
8. **The Conclusion: Innovation and Self-Different Fractals** **58**

References **71**

Index **89**

1

Introduction: Self-Different Fractals and Innovation

The ideas that are being presented here carry the vision of developing a ***"Manifesto for Higher Education Governance"****.*

From a philosophical perspective, our focus is on exploring how a "philosophy of dialectics", a "dialectics of innovation", and a "self-different fractal system" of innovation can be interconnected and translated into one another.

Paraphrased, our research question is: *How to apply the concept of a "self-different fractal innovation system" in a way, so as to describe how governance in higher education can interlink the concepts of "entrepreneurial university", "academic firm" and "epistemic governance"?* These concepts are of relevance for research (knowledge creation and production) and innovation (knowledge application) in higher education.

Our research question is explorative, and its application is to a certain extent experimental. However, the identified concepts (entrepreneurial university, academic firm, and epistemic governance) were selected, as these appear to be of relevance for the contemporary *transition in transformation* of the unfolding higher education systems. We review some of the important trends that occur, and reflect these further in context of the current discourses and narratives. While ideas about the "entrepreneurial university" already express some prominence in the debates, the "academic firm" may still be more of a speculation, in the sense that the formation of academic firms is consistent with what modern theories are saying about knowledge and innovation; furthermore, academic firms are designed as enablers of a productive interaction with entrepreneurial universities for purposes of a co-creation and co-evolution of knowledge and innovation. But the question arises whether academic firms truly exist or if they are just exceptions in the organization of mod ern business. We do not offer a conclusion to this important question at the moment. Because even if academic firms represent a rarity,

DOI: 10.1201/9781315209234-1

at least, "according to theory", they have the potential to become more important in the future. Again, academic firms do not necessarily have to operate as completely distinct entities. Aspects of academic firms could be embedded as specific characteristics within the context of traditional "commercial firms".

Perhaps the situation is for the governance of higher education similar. Possibly New Public Management (in variations) represents the dominant mode of governance in higher education. Nevertheless, there is also a certain disenchantment and disillusionment with New Public Management (NPM), as it cannot address sufficiently all the relevant aspects of higher education. Network structures and network processes underscore a complementary need for network governance. *In addition, epistemic governance enhances the ability of higher education institutions to combine and integrate different knowledge and innovation modes, thus promoting higher education institutions in their capacity of a knowledge and innovation development.* All of this feeds into a "pluralism of governance" in higher education.

We start our analysis with a short "philosophical tractate" on self-different fractals and the dialectics of innovation. but the focus of analysis refers to what is (assumably) occurring and taking place in the empirical world. Still, the "constructed point of departure" is that it is the "ideas" that are driving the empirical analysis. "Not everything is new under the sun". Therefore, for the review of discourses about governance of higher education, our interest was to reconstruct discourses of the last years, so to have here a somewhat broader perspective.

Already some time ago we applied ideas about fractals to higher education. The goal was to present the concept of FREIE, which serves as an acronym representing specific principles: *Fractal Research, Education and Innovation Ecosystem* (Carayannis and Campbell, 2011). There was a nuanced element of wordplay involved. In the German language FREIE or "Freie" means "the free", or "free persons", "free people". *So, the intention here was to underscore a so-called education-democracy-nexus, implying that higher education also should promote democracy enhancement. Therefore, higher education governance has a responsibility for democracy development.* The concepts of Third Mission (and Forth Mission), as well as their governance, address this interaction of higher education and democracy.

It began with FREIE more than ten years ago.

This time, we revisited the "world of fractals", exploring them through more radical philosophical ideas and approaching them in an unconventional way.

Fractals are typically perceived as being "self-similar". In a philosophical extension and stretch of this, we now propose to introduce the "self-different" fractals. This work aims to translate the traditional "philosophy of dialectics" into the modern concept of self-different fractals, *bridging and inter-bridging classical philosophical ideas with contemporary theories of innovation and innovation systems.*

Figure A illustrates self-similar and self-different fractals in a simple graphical image, while Jürgen Pirecki-Giefing presents a more artistic interpretation of self-similar fractals in Cluster 1.

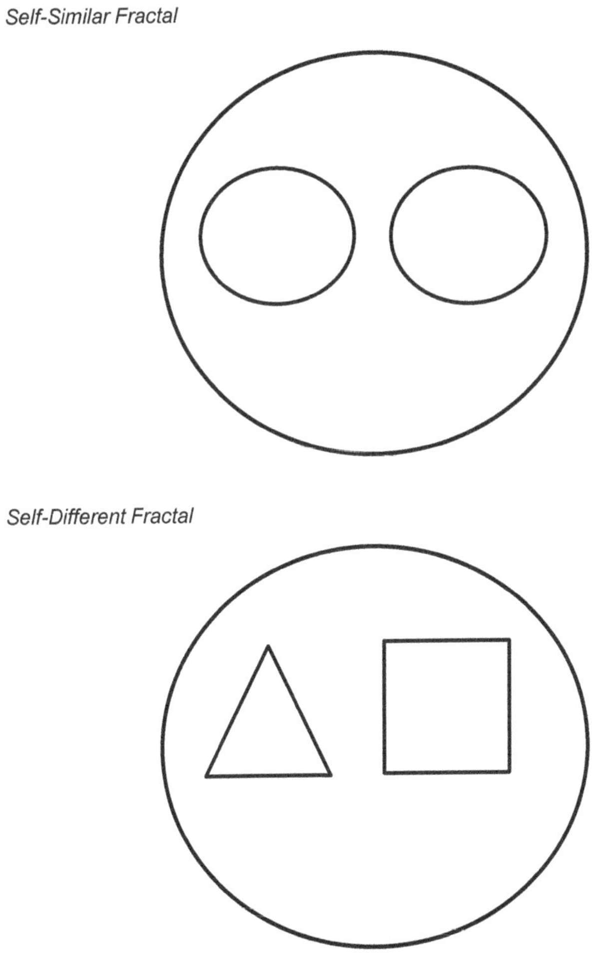

Fig. A Self-Similar and Self-Different Fractals.
Source: Authors' own conceptualization and visualization.

The interaction between "thesis" and "antithesis" may be disruptive, but with proper management and governance, it can drive creativity, co-creation, collaboration, innovation, co-development and development. *"Self-different fractal systems", which operate through multiple modes, have a higher potential for innovation. The co-modal ("thesis", "antithesis ") or multi-modal nature of self-different fractal systems enhances their "innovative innovation" capabilities.* The interplay of different modes can inspire creativity, resulting in new ways for creating, producing, and applying knowledge.

Effectively governing innovation may involve designing governance structures that can manage self-different fractal systems of innovation. Effective governance of a self-different fractal innovation system fosters the evolution of knowledge, advancing it to new stages. Drawing from a "dialectics of innovation", governance in these complex ecosystems allows for the creation of a synthesis, where diverse knowledge and innovation paradigms and modes converge to form new, productive structures, processes, and networks.

Our "conceptual experiment", being presented here, focuses on innovation and higher education systems by proposing to understand the (1) entrepreneurial university as the "thesis", (2) the academic firm as the "antithesis", (3) and epistemic governance as the governance approach of "synthesis".

This monograph is organized into several sections, as is outlined in this brief preview:

In Chapter 2, there is a first leading-in to the topic of governance of and in higher education.

Chapter 3 elaborates on the "Knowledge State".

Chapter 4 outlines a synopsis of the concept of "Epistemic Governance" in higher education.

In Chapter 5, different knowledge concepts are being reviewed, such as "innovation", "Mode 1" and "Mode 2", and characteristics of a "Mode 3" university (higher education institution) are explained in more detail.

Chapter 6 portrays new academic careers.

Chapter 7 provides a more detailed exploration of the concept of the "Academic Firm".

Chapter 8, the conclusion, discusses and summarizes the concept of "Self-Different" fractal innovation systems, while also tying together the arguments related to the entrepreneurial university, academic firm, and epistemic governance. The analytical flow of arguments finally is concluded in a "Conceptual Glossary" at the end of chapter.

Self-Similar Fractals: Images by Jürgen Pirecki-Giefing.

Fig. Cluster 1 Images of "Self-Similar" Fractals
Author & Creator: Jürgen Pirecki-Giefing.

2

Governance of and in Higher Education

In a historical understanding, governance associated closely with steering, which originally also could have meant to steer machines or ships.

> "In etymological terms, the origin of the word 'governance' derives from the ancient Greek verb kybernein (infinitive) or kybernao (first person) that meant steering, guiding or maneuvering a ship or a land-based vehicle, and was used the first time metaphorically by Plato for depicting the governing of men or people (people would be here the modern application)" (Campbell and Carayannis, 2013, p. 3).

The term "steering" is also linked to the prefix "cyber", as seen in words like "cybernetics". In modern English, "governance" is connected to terms like "government" and "govern" (European Commission, 2001a; 2001b). Cybernetics focuses on feedback and regulatory systems, as noted by scholars such as Wiener and Umpleby (Wiener, 1948; Umpleby, 1989; Müller, 2018).

> "Should this close link between government and governance be continued conceptually, then a definition of governance may be: governance describes how a government governs. One may also say, alternatively: governance addresses how a government governs based on feedback. 'Cybernetic governance' may be paraphrased as a Governance of Governance" (Campbell and Carayannis, 2013, p. 3; Carayannis and Campbell, 2019; Carayannis et al., 2018).

Governance, steering, and control share similar ideas, but differ in scope. When analyzing society, governance is the broadest concept,

DOI: 10.1201/9781315209234-2

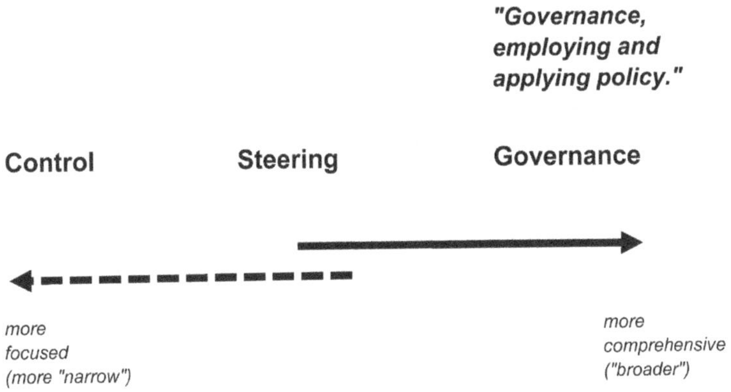

Fig. B The Two Poles and the Continuum of Control and Governance.
Source: Authors' own conceptualization and visualization based on Campbell & Carayannis (2013) and Carayannis & Campbell (2021).

encompassing more than steering, which in turn covers more ground than control. Governance can involve steering goals but is not limited to it, while steering may aim for control. However, controlling society or advanced economies is questionable in terms of feasibility and desirability. After the spread of market economies and the fall of Eastern European Soviet-style communist regimes, doubt grew about the possibility of societal control. Control works well for machines, but social control would need to rely on feedback mechanisms (see Figure B for a visualization).

A possible functional definition of governments or the broader political system could be proposed based on this understanding: "*... government governs or wants to govern society. With policy the government governs (steers) or intends to govern (to steer) society and the different systems (subsystems) of society, including the economy and the higher education sector*" (Campbell and Carayannis, 2013, p. 14). This governance function of governments can also be seen as a "self-rationale of the political system" (Carayannis and Campbell, 2006a, p. 9): "The political system has or should express a responsibility for the overall performance of a society. The 'governance of society' can be defined as a self-rationale of politics: through policy (policy-making) and legislation or – alternatively – steering, coordination and communication the political system attempts to influence the dynamics of a society and economy, and tries to support the performance (and self-rationales) of the other systems (Campbell, 2001, p. 428). Put in summary, the political system is interested in effectively stimulating and coordinating the performances of the other systems and thus enhancing a synergetic performance surplus."

The concept of the "political economy" focuses and refers to state-economy interactions, but also frames this in context of society. Michael

J. Sodaro (2004, p. 297) presents this in the following way: "Political economy is the study of how communities pursue collective economic goals and deal with conflicts over resources and other economic factors in an authoritative way by means of government". Sodaro (2004, p. 308) incorporates this definition into his broader definition of welfare state: "Broadly defined, the welfare state is a form of political economy in which the state assumes responsibility for the general welfare of its population, especially its most vulnerable elements, through spending on such items as education, housing, health care, pensions, unemployment compensation, food subsidies, family allowances, and other programs." It should be emphasized that Sodaro qualifies education as an element belonging to the welfare state.

Governance and our understanding of it have grown significantly more complex in recent years (Elfert and Ydesen, 2024; Elken, (2024); Kallo and Välimaa, 2024; Treib et al., 2007). R.A.W. Rhodes (1996) formulates the question, whether governance can exist without the need for formal government structures.[1] He puts it as: *Governing without Government?* Rhodes (1996, p. 653) distinguishes between different "uses of governance": "as the minimal state"; "as corporate governance"; "as the new public management"; "as good governance"; "as a socio-cybernetic system"; and as "self-organizing networks". "Initially, the 'new public management' (NPM) had two meanings: managerialism and the new institutional economics. Managerialism refers to introducing private sector management methods to the public sector. ...The new institutional economics refers to introducing incentive structures (such as market competition) into public provision".[2] Rhodes (1996, p. 652) emphasizes particularly the self-organizing networks: "I stipulate that governance refers to 'self-organizing, interorganizational networks' and argue these networks complement markets and hierarchies as governing structures for authoritatively allocating resources and exercising control and coordination ". As Rhodes (1996, p. 658) says, "Governance is about managing networks", and "I use the term network to describe the several interdependent actors involved in delivering services" (Rhodes, 1996. p. 658). Characteristics in common of this network-style governance are for Rhodes (1996, p. 660): "Interdependence between organizations"; "Continuing interactions between network members"; "Game-like interactions"; and: "A significant degree of autonomy from the state. Networks are not accountable to the state; they are self-organizing".[3]

[1] See also Neave (1988a and 1988b, 2007) and Neave et al. (1991).

[2] Rhodes (1996) created an intellectual impact and influenced the "Network Governance" narrative of Ferlie et al. (2009).

[3] Furthermore, see Rhodes (2008), but see also Rhodes et al. (2009).

Some years before Rhodes, already Charles Raab (1994, p. 17) posed: "New forms of governance constitute attempts to cope with the complexity, dynamics and diversity of socio-political systems through more effective instruments than were disposed of by traditional mechanisms of state control or the market. These instruments may take the form of co-ordinated action in networks linking many organizations, some of which are self-regulating, and in which governance may steer 'at a distance' … but may not necessarily control or determine outcomes". A new paradigm for public governance of higher education in the Netherlands (Kickert, 1995)[4] follows (followed) the logic of a *steering at a distance*. Geuna et al. (2003) posed the question, whether governance of and in higher education should see science more as a *public good* (science-as-a-public-good) or, alternatively, as a *network* (science-as-a-network). Of course we may ask: Is there a difference between networks and systems, or *do systems resemble networks?*[5]

Advanced societies and economies often are being described as knowledge societies, knowledge economies, but also as *knowledge democracies* (Veld, 2010). Knowledge and innovation (the application of knowledge) act as crucial drivers and are themselves becoming more complex (Dubina et al., 2012). Industry 4.0 and 5.0 (*Industry 4.0, Industry 5.0*) emphasize the growing importance of knowledge and innovation in industry (Carayannis et al., 2022a; Carayannis and Morawska-Jancelewicz, 2022; Campbell and Carayannis, 2023b and 2023c). Higher education and its massification also play a crucial role (OECD, 2024). In reference to knowledge at HEIs (higher education institutions) or in the context of higher education systems, concepts were developed such as "Mode 2" (Gibbons et al., 1994) and "Mode 3" knowledge production (Carayannis and Campbell, 2009), "Triple Helix" (Etzkowitz and Leydesdorff, 2000; Leydesdorff, 2012; Leydesdorff and Smith, 2021; Cai and Lattu, 2022) or "Quadruple and Quintuple Helix" innovation systems (Carayannis and Campbell, 2009, 2010, 2017, 2019, 2020, and 2022; Carayannis et al., 2012, 2021, and 2022b). The "academic firm" (Campbell and Carayannis, 2016a and 2023a; Campbell et al, 2013; Campbell and Güttel, 2005), *and furthermore the notion of quality of democracy as an innovation enabler* (Campbell, 2019; Campbell et al., 2015).

Higher education, higher education institutions and organizations and higher education systems are central for knowledge societies,

---

4 Raab (1994, pp. 17, 20) mentions how Walter Kickert presented first his 1995 article as a paper for a workshop that was organized by the European Consortium for Political Research at the University of Essex back in 1991.

5 In this respect, see, for example, the discussion in Kritzinger et al. (2006).

knowledge economies and knowledge democracies. This is true for the Global North, but is equally true for the Global South. Since knowledge (knowledge production, knowledge creation interlinks with innovation (knowledge application), higher education plays furthermore a decisive role for innovation and the different domains of innovation (Carayannis and Campbell, 2021*). For the governance of higher education results, therefore, the further challenge, how to link knowledge production (creation) and innovation in the higher education system with knowledge and innovation in society, economy and democracy* (Burkert et al., 2024; Gepert et al., 2024).

The functional profile and goals of universities and other higher education institutions (HEIs) of the higher education sector, *how can this be described, defined or explained* (see Figure C)?

"Traditionally, higher education focuses on teaching and research. The Humboldtian concept of unity of research and teaching 'highlights the essential interconnections between these activities. Universities are involved in both teaching and research, primarily focusing on basic research (see, for example, The National Science Board, 2010, p. 15 [Chapter 4]). In recent years discussions have emerged on whether the core functions of teaching and research still adequately describe the higher education sector, or if 'third-mission' activities"[6] should also be considered (Campbell and Carayannis, 2013, p. 5).Within this framework, research and teaching are seen as the fundamental activities of universities (higher education institutions) (Campbell, 2013a; Gottweis, 1998; Schibany and Gassler, 2010).

"Can academia encompass a third mission of economic development in addition to research and teaching? Etzkowitz and Leydesdorff (2000, p. 110) argue that a 'second academic revolution' has been underway since World War II, becoming more visible since the end of the Cold War". Universities and the higher education sector are now expected to generate knowledge that benefits societal and democratic development while being economically sustainable. This role positions these institutions as essential drivers of progress in societies that rely on knowledge as a foundation. Emphasis has shifted toward "knowledge societies",

---

[6] Pasternack (2008) argues that Wilhelm von Humboldt never explicitly used the phrase "unity of research and teaching". Instead, this phrase emerged later as an interpretation of Humboldt's educational ideas. Humboldt viewed sciences as a continuous process of inquiry rather than a fixed body of knowledge and differentiated between the roles of schools and universities, seeing universities as spaces for advanced, research-oriented learning. "According to Pasternack, Humboldt emphasized two aspects: defining sciences as an ongoing research process and distinguishing between teaching at schools and universities" (Campbell and Carayannis, 2013, p. 5).

"knowledge economies", and "knowledge democracies", underscoring the importance of knowledge in these areas. Higher education thus plays a critical role in fostering both intellectual and economic growth in modern society (Carayannis and Campbell, 2011; 2012, p. 55). Bengt-Åke Lundvall (1992, p. 1) emphasized that most fundamental resource in the modern economy is knowledge, making the most important process is learning.He noted that the unique nature of knowledge renders standard economics less relevant."

Higher education institutions are increasingly linking *knowledge creation and production (research)* with its application and diffusion, aligning with the concept of innovation, which *emphasizes knowledge application* (Carayannis and Campbell, 2010, p. 45). This integration highlights the role of universities beyond teaching and research, encompassing a "third mission" focused on real-world impact (Pausits, 2015; see also Peschke, 2021, and Peschke et al., 2023). Innovation exemplifies these expanded activities, underscoring the importance of HEIs in driving societal and economic advancement.

The concept of the "Entrepreneurial University" (Etzkowitz, 2003) addresses these issues. Etzkowitz and Leydesdorff (2000, p. 111) describe the Triple Helix model of knowledge and innovation based on "University-Industry-Government Relations". The Quadruple Helix adds the perspectives of a "media-based and culture-based public" and of "civil society" (Carayannis and Campbell, 2009, p. 207; 2011, 2012; pp. 13-14; 2017). The Quintuple Helix frames the Quadruple Helix within the "natural environments" of society (Carayannis and Campbell, 2010, p. 62; 2019; 2020; Carayannis et al, 2012, 2021, and 2022b; Campbell and Campbell, 2011).[7] Other examples of third-mission activities of HEIs include the *"civic mission"* of "civic education" or *"democracy education"*.[8] *"Community Engagement"* is another illustration for third mission (European Commission, 2024). Additionally, the concept of the "Responsible University" should be addressed (Sørensen et al., 2019).

While uncertain, some speculate that a fourth function for higher education may emerge beyond current frameworks. Reflecting on this possibility requires considering the evolving role of universities in society. Continual reflection on the "future of the university" is crucial to anticipate new functions and keep higher education adaptive to societal changes. Scholars like Mittelstrass (2010), Burkert (2021), and Martin (2003) have already reflected on this.

---

7 For reflections about a possible "Quattro Helix Innovation" system see furthermore Danilda et al., 2009.

8 Interesting in this context is the research work of Professor Jasminka Ledić and of her Croation research team in Croatia.

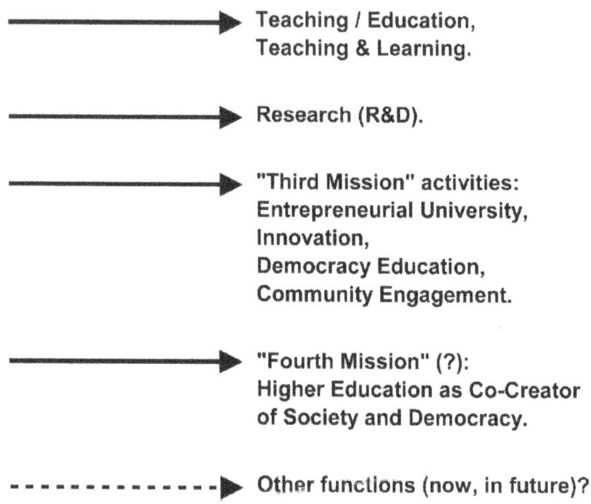

Fig. C Functional Profile of Universities (HEIs).
Source: Authors' own conceptualization and visualization based on Campbell & Carayannis (2013) and Carayannis & Campbell (2021).

Third mission activities and interdisciplinary (multidisciplinary, transdisciplinary) research are crucial for knowledge production and innovation at universities. Interdisciplinarity aims to integrate diverse research approaches for better comprehension of complexity (Wagner et al. 2011), while third mission activities focus on enhancing innovation. This clearly expresses characteristics of an "entrepreneurial university", networking with society, democracy and the economy including various types of companies and academic firms. Certain cross-associations between third mission and interdisciplinarity are assumed, though empirical verification of this hypothesis is still needed.

Compagnuccia and Spigarelli (2020) conducted a systematic literature review on the Third Mission, stating:

"To the best of our knowledge, no systematic literature review has been performed on the TM to comprehensively explore its heterogeneous functions, constraints, clashes and incorporation within education and research."[9]

"They examined 134 peer-reviewed articles which were published between 2004 and May 2019 providing both a descriptive and thematic analysis."[10]

---

[9] Compagnuccia and Spigarelli, 2020, p. 1.
[10] Compagnuccia and Spigarelli, 2020, p. 1.

They formulated the following key statements:

1. The Third Mission (TM) is broadly understood as "a contribution to society"[11].
2. The TM is a multidisciplinary, complex, and evolving phenomenon linked to the broad social and economic mission of Universities.[12]
3. The paper reveals the potential and constraints of the recurring themes of the TM, with a focus on engaging non-academic stakeholders. It also suggests recommendations for scholars and policymakers. [13]

In their key conclusion, the authors argue:

1. At the core of the Third Mission is the concept of the "entrepreneurial university" which serves as a model for the TM of universities[14].
2. Third Mission implies that universities engage in knowledge transfer and entrepreneurship education, emphasizing these functions within the TM framework.[15]
3. Wider forms of engagement will become increasingly important (Engagement of academics and external stakeholders[16]). Academics (faculty and staff) should be considered within a broader context.
4. New approaches are necessary for measuring and evaluating the Third Mission, which differ from those used for teaching and research. Additionally, a diverse range of evaluations is needed, as "a one-size-fits-all model for evaluating TM performance and its impact cannot be applied universally.[17]
5. Implementing the Third Mission requires new or complementary organizational approaches in strategy and governance (Strategic orientation of TM[18]).

Wagner (et al., 2011) conducted a literature review to identify the common understandings of *multidisciplinarity, interdisciplinarity and transdisciplinarity.*

1. Multidisciplinarity:
    1.1 Multidisciplinary approaches combine different disciplinary and professional perspectives, adding breadth and expanding available knowledge, information, and methods.[19]

---

[11] Compagnuccia and Spigarelli, 2020, p. 1.
[12] Compagnuccia and Spigarelli, 2020, p. 1.
[13] Compagnuccia and Spigarelli, 2020, p. 1.
[14] Compagnuccia and Spigarelli, 2020, p. 20.
[15] Compagnuccia and Spigarelli, 2020, p. 20.
[16] Compagnuccia and Spigarelli, 2020, p. 20.
[17] Compagnuccia and Spigarelli, 2020, p. 20.
[18] Compagnuccia and Spigarelli, 2020, p. 20.
[19] Wagner et al., 2011, p. 16.

  1.2 They speak as separate voices, in encyclopedic alignment, an ad hoc mix, or a mélange. Disciplinary elements retain their original identity.[20]
2. Interdisciplinarity:
  2.1 Interdisciplinary approaches combine separate disciplinary data, methods, tools, concepts, and theories to create a holistic view or common understanding of complex issues.[21]
  2.2 "Critical indicators of interdisciplinarity in research include evidence that the integrative synthesis is different from, and greater than the sum of its parts.[22]
3. Transdisciplinarity:
  3.1 More recently, the term has referred to a new mode of knowledge production that draws on expertise from various organizations and collaborative partnerships for sustainability. This integrates research from different disciplines with the stakeholders' knowledge in society, producing a transdisciplinary outcome greater than the sum of its parts,with a more comprehensive scope and diverse components.[23]

In summary, we can conclude the following:

1. *Multidisciplinarity:*
  1.1 Indicates a cooperation of disciplines.
2. *Interdisciplinarity:*
  2.1 Refers to issues and problems too complex for a single discipline to solve.
  2.2 Of course, we always can ask: Which issues can be tackled by only one discipline?
3. *Transdisciplinarity:*
  3.1 The difference with interdisciplinarity is not so much in the "what" of knowledge, but the "how" of knowledge production;
  3.2 Cooperation between academic research communities and practitioners is crucial for effective knowledge production;
  3.3 It impacts research design by encouraging scholars to go beyond purely academic spheres ;
  3.4 Radically new knowledge can emerge from the interaction of scientific and artistic research (Carayannis and Campbell, 2015).

---

20 Wagner et al., 2011, p. 16.
21 Wagner et al., 2011, p. 16.
22 Wagner et al., 2011, p. 16.
23 Wagner et al., 2011, p. 16.

Figure D illustrates the relationship between internal governance (institutional self-governance within universities) and external governance (influence from other institutions or systems). It frames system governance *within the poles of "internal" and "external" self-governance*, showing the impact of external governance efforts on internal governance. Within systems like higher education, there is a dynamic interplay between internal and external governance. This can be seen as self-governance at a systemic level, influenced by external governance from other systems or sectors.[24]

A system or sector, like *higher education,* shows the *interplay of internal governance (institutional self-governance) and external governance. When this interaction occurs* within the boundaries of systems, it can be seen as self-governance at the systemic macro-level. However, external governance from other systems or sectors can also influence it (Campbell and Carayannis, 2013, p. 20).

Traditionally, governance was linked to governments. Recently,the concept has expanded to include non-governmental forms. Initially, governance in higher education was seen as governing higher education. However, there's been a shift towards recognizing self-governance within higher education systems, leading to a combined approach of governance within higher education. The interaction between external and internal governance in higher education presents significant challenges. These are characterized by complex networks spanning institutions and sectors, involving both cooperation and competition, often described as *"co-opetition"* (Brandenburger and Nalebuff, 1997). R.A.W. Rhodes 1996, p. 658)[25] uses the metaphor "Governance as Self-organizing Networks" (Goodwin, 2009).

With a growing awareness of self-governance and internal governance, governance of higher education has gradually shifted to a *combined governance of and in higher education. The interplay and interfaces of external and internal governance* are important but also present significant challenges for higher education (Campbell and Carayannis, 2013; p. 21).

---

[24] In systems theory, "external governance" resembles an "external environment" for organizations, sectors and systems. These entities must reflect and adapt to their external social, societal, political, economic and ecological environments, including external governance attempts and measures (see Figure D, and references : Willke, 1983, 1989, 1998; Luhmann, 1988).

[25] Compare also with Rhodes (2008) and Rhodes et al., 2009.

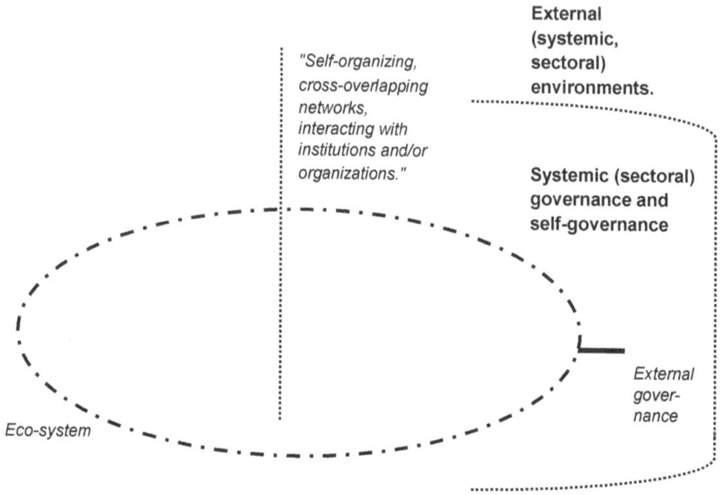

Fig. D Systemic Governance and "Internal" and "External" Self-Governance.
Source: Authors' own conceptualization and visualization based on Campbell & Carayannis (2013) and Carayannis & Campbell (2021).

The future of universities and higher education institutions is intertwined with the future of governance within this sector. Governance trends reflect the future of higher education (Campbell, 2024; Campbell and Stransky-Can, 2020; Campbell et al., 2021; Faulhammer and Sagl, 2024; Stachowiak-Kudła et al., 2023; Westa, 2024; Ziegele, 2024). Governance of higher education should be guided by visions of *future innovations* (Campbell and Bösch, 2020; Weber, 2024). Effective higher education governance can balance crisis impacts (Pausits et al., 2023). Speculating on the future requires researching the past of higher education systems (Welp-Park, 2023). Governance *transformation and the governance of transformation are interconnected* (Burkert et al., 2024). "Systemic governance" of higher education systems marks a new phase of importance.[26] Effective higher education governance can also balance crisis impacts (Pausits et al., 2023).

[26] At the University for Continuing Education Krems (in Austria) the *"Center for Higher Education Governance and Transformation"* (https://www.donau-uni.ac.at/en/university/faculties/education-arts-architecture/departments/higher-education-research/centers/higher-education-governance.html) was established. This center is devoted to research and to reflect on *governance of and in higher education.*

3

The "Knowledge State": Governance of Higher Education in the Knowledge Society, Knowledge Economy and Knowledge Democracy

Theories of the knowledge-based society and economy emphasize the growing role of knowledge in driving economic and societal progress (Rodrigues, 2002; Carayannis and Campbell, 2009 and 2021). Scholars like Lundvall highlight that knowledge is the key resource in modern economies, with learning being the essential process. Lundvall (1992, p. 1) states, "the most fundamental resource in the modern economy is knowledge" and "the most important process is learning".[27] This shift from a knowledge-based economy to a knowledge economy, and from a knowledge-based society to a knowledge society, underscores the central role of knowledge in these contexts. Additionally, the concept of a knowledge democracy is emerging, highlighting the democratic dissemination and utilization of knowledge within society (Campbell, 2019; Campbell and Carayannis, 2023a, 2024b and 2024c; Carayannis and Campbell, 2006a, p. 19; 2009, p. 224; Carayannis and Campbell, 2010, p. 52; 2012, p. 55; 2014 and 2021; Carayannis et al., 2022a and 2022b).

The concept of political economy explores the links between politics, society, and the economy. Previously, we discussed political economy

[27] Lundvall (1992, p. 18) highlights knowledge as a crucial element in the knowledge economy, emphasizing its essential role in economic processes and innovation: Knowledge does not decrease in value when used. On the contrary, its use increases its value. Knowledge is not scarce like natural resources and technical artifacts. Some knowledge can be easily transferred between economic agents while other knowledge is tacit and embedded in individual, or collective, agents.

DOI: 10.1201/9781315209234-3

and the welfare state, as defined by Michael J. Sodaro. Sodaro (2004, p. 308) considers the welfare state a type of political economy, in that sense *the welfare state represents a form of a political economy*. In the context of the knowledge economy and society, political economy of the welfare state faces new challenges and opportunities to develop a clearer profile.[28]

Framed within knowledge society and economy, political economy should focus on how the state promotes and utilizes knowledge, including research, education, and innovation, to enhance societal well-being and economic performance.Campbell and Carayannis (2013, p. 22) emphasize that *for the knowledge society and economy the political economy should stress the state's role in leveraging knowledge and innovation for societal welfare and economic performance* (Campbell and Carayannis, 2013, p. 22).

The concept of the "Knowledge State" (Campbell, 2006a, pp. 25-27; Campbell, 2006b; Duberry, 2022) suggests that governments play a key role in governing society and the economy with a focus on knowledge. This perspective defines political systems and governments based on their functional interests. The complexity of modern society involves governance intertwined with self-organization and self-steering, including self-governance. The "innovation system" overlaps with political, economic, research (R&D), education, and higher education systems (HE sector) (Carayannis and Campbell, 2006a, p. 17).

The "knowledge state" concept, as described by Campbell (2006a, p. 26) has key attributes: 1) Politics acknowledges the key role of knowledge in society, the economy, and democracy. 2) Political entities aim to foster various forms of knowledge and knowledge systems. 3) Policies focused on knowledge and innovation, seek to bolster economic development by leveraging the knowledge base of society for the benefit of the economy. However, it is also imperative that economic progress contributes to the sustainable development of the knowledge base, thereby maintaining a balance between the interests of society, the economy, and democracy. The knowledge states also can be understood as an "enabling state" (Page and Wright, 2007).

---

[28] Political economy and the knowledge economy can be seen as opposing concepts. Loet Leydesdorff (2012, pp. 25, 32) supports this by highlighting their contrasting aspects: "In a knowledge and innovation aim to enhance economic development. ... The shift from a political economy to a knowledge-based economy driven by techno-scientific advancements became a major competitive force at the macro-level after the fall of the Berlin Wall and the Soviet Union's collapse (1990-1991). The political economy transformed into a knowledge-based economy as the battle between different political economy models became obsolete". Instead, we see the knowledge economy as providing a specific context for the development of political economy.(Campbell and Carayannis, 2013, p. 22).

Pechar and Andres (2011, p. 25) establish a direct connection between welfare regimes and higher education. They assert that all OECD countries have seen significant growth in higher education since the latter half of the twentieth century, primarily due to its integration into national welfare policies. However, the importance placed on education, particularly higher education, within the broader framework of welfare policies varies among OECD countries. To understand these differences in national approaches to higher education participation, funding, tuition, and student financial aid, Pechar and Andres employ the concepts of "welfare regime" and a "trade-off" hypothesis.

The term "welfare regime" is essentially interchangeable with "welfare state". This conceptual link between welfare state, welfare regime, political economy, and higher education policies holds significant relevance for governance, particularly in the governance of higher education. Pechar and Andres (2011, pp. 27-28) refer to the typology-of-welfare-regimes of Gøsta Esping-Andersen (1990, p. 26) for practical and data-based testing of various hypotheses. The approach of Esping-Andersen involves identifying different arrangements between the state, market, and family in terms of social rights and welfare-state stratification, which are clustered into distinct regime types rather than being linearly distributed.

Esping-Andersen (1990, p. 37) introduces the concept of "de-commodification" as a way to categorize different welfare-state regimes: "The variability of welfare-state evolution reflects competing responses to pressures for de-commodification. Rather, the concept refers to the degree to which individuals, or families, can uphold a socially acceptable standard of living independently of market participation." Esping-Andersen (1990, pp. 50-54) identifies three types of welfare regimes (see also Pechar and Andres, 2011): *liberal welfare regimes* (for example Canada and the U.S., Australia and New Zealand, the UK); *conservative welfare regimes* (for example Austria, France, Germany, the Netherlands and Belgium, Switzerland, Italy); and *social-democratic (universal) welfare regimes* (for example Norway, Sweden, Denmark, and Finland). Expanding this typology globally could reveal whether additional welfare regimes beyond these are necessary.

In the innovation system, knowledge and its application for problem-solving play a central role (Carayannis and Campbell, 2010. p. 45; 2021). This system intersects with various other societal spheres, forming a complex interface between politics, society, and the economy. This interconnection offers new opportunities for the political system, allowing it to govern and steer not only the innovation system but also other sectors of society. Through the innovation system, the political system can exert influence and attempt to govern or guide other societal systems. "*Via the innovation system, the political system can attempt to govern and steer the other sectors of society*" (Campbell and Carayannis, 2013, p. 24).

In a political-economic regime, policies must directly or indirectly focus on knowledge, considering the implications of the knowledge society and economy. Innovation systems enable political governance by influencing the economic system through innovation policy, emphasizing the importance of knowledge. This positions the innovation system as a key interface between politics and the economy. Additionally, the knowledge-based economy suggests that innovation policy may often be more effective in enhancing economic performance than traditional economic policies.

"Innovation policy, which recognizes the importance of knowledge, allows the political system to indirectly influence the economic system. This highlights the innovation system as a key interface between politics and the economy. The knowledge-based economy and society suggest that innovation policy can often be more effective than traditional economic policy in supporting economic performance. In advanced societies the overlap of the innovation system with politics and the economy becomes increasingly important. Therefore, for knowledge-based economies and societies innovation systems and policies are crucial areas for analysis in Political Economy or International Political Economy" (Carayannis and Campbell, 2006a, pp. 18-19).

"Innovation policies can be seen as an evolution beyond traditional economic policies, indicating a maturation of governance in the political economy (Campbell & Carayannis, 2013, p. 24). These policies allow governments to govern various knowledge-based systems, extending beyond the economy. The knowledge aspect serves as a common thread, offering a comparative basis for policies across different fields, as it is fundamental to all innovation policies.

The higher education sector, including universities, is becoming increasingly crucial for the knowledge economy, society, and democracy. These concepts rely heavily on high-quality knowledge. Universities and higher education institutions generate this essential knowledge through tertiary education, research (including basic, interdisciplinary and transdisciplinary research), and innovation. This knowledge is made accessible to the economy, society, and democracy, contributing to their development and advancement. Etzkowitz and Leydesdorff (2000, p. 109) highlight that the "Triple Helix thesis states that the university can play an enhanced role in innovation in increasingly knowledge-based societies" (p. 109).[29]

Governance in higher education must consider the entire functional profile of universities (higher education institutions) into account. This includes governance of university research, teaching[30], and Third Mission activities of the higher education sectors within the higher education system.

The increasing importance of higher education in advancing knowledge economies, societies, and democracies highlights the need for effective governance. As higher education becomes essential for knowledge-based economies and societies, political systems and governments must prioritize its governance. *"This growing importance of higher education for advancing knowledge economies, societies and democracies, underscores the need for effective governance,both conceptually and in policy"* (Campbell and Carayannis, 2013, p. 25; see furthermore Campbell, 2024; Campbell and Stransky-Can, 2020; Campbell et al., 2021; Burkert et al., 2016 and 2024).

Governance of society and the economy through innovation policies partially intersects with higher education policies (Larédo and Mustar, 2001). Advanced innovation policy relies on a mature or evolving higher education policy making higher education policies more prominent on government agendas. Given higher education's significance, it represents a system where governments frequently implement new policies, *likely becoming a greater priority for decision-makers.*

Jeroen Huisman (2009, p. 3) emphasizes that higher education, like many societal sectors, has experienced shifts in governance. In many Western countries, higher education has been a leader in experimenting with new steering approaches often resulting in less governmental interference or a shift from *ex ante* control to *ex post* evaluation, granting more autonomy to higher education institutions".

Huisman (2009, pp. 3–4) introduces the concept of "The Invisible Hand of Governance," highlighting governance's subtle influence on higher education systems. He identifies several trends : 1) Variations in the speed and focus of governance policies among higher education systems.

---

[29] Advancement in the knowledge economy and knowledge society (within the contextualization of the twenty-first century, but also beyond), both now and in the future, relies heavily on the continued development of higher education, regardless of how it evolves.

[30] As Magdalena Fellner (2025, p. 319) states: "The interests of powerful groups inevitably shape the access regulations at higher education institutions. Through governance mechanisms, these groups can exert further control over the student population. These mechanisms are contingent upon specific historical and contextual factors".

2) A balancing act between market forces and evaluative states or distant steering. 3) Increasing interplay between domestic and supranational policies and steering approaches. These trends may lead to the emergence of "multi-level governance" in higher education systems (Hooghe and Marks, 2001).

Box A examines the characteristics of political governance focusing on the public promotion and support of university research (R&D)[31] in sciences and arts within the higher education system.[32] This simplified visualization, requires further elaboration for a more detailed discussion and assessment.[33]

Box A: Political Governance (Steering) through the Public Promotion of University Research.

Analyzed systems:	*Political system and research (R&D) in the higher education system.*
Research (R&D) in the higher education system:	
"Elements"/parts:	E.g., university R&D in the sciences and arts (e.g., basic research).
"Self-rationale" of the self-organization of the higher education system:	In general: e.g., increase of effectiveness (including quality) of R&D.
	In particular: e.g., increase of the effectiveness of university research across the following "dimensions" in the sciences and the arts: Quality, efficiency, relevance, and viability (sustainability).
Political system:	
"Elements"/parts:	E.g., government or the "state" (the "knowledge state" in the "knowledge democracy").
"Self-rationale" of the self-organization of the political system:	In general: e.g., political governance (steering) of society or the subsystems of society, such as the sciences and arts in the higher education system.
	In particular (a): e.g., support of the development of advanced knowledge societies by applying R&D, educational, technology, and innovation policies (also promoting research).
	In particular (b): e.g., public promotion of research (R&D) in the sciences and arts in the higher education system (university research = "primary location within society" for basic research?).

Contd.

[31] On the importance of technology for innovation (and innovation governance) see Hanschitz (2023).

[32] See also Biegelbauer (2010).

[33] Furthermore, see Figure F in Chapter 4.

Systems theoretical implications:	Translating the "self-rationale(s)" of the political system into the "self-rationale(s)" of higher education, whithout malfunctioning the "autonomy" and "self-rationale(s)" of the higher education system.
Governance (steering) options for politics:	E.g., allocation and re-allocation of (public) resources. E.g., persuasion through (mutual) communication and the designing and development of incentives (for example, the allocation of resources, the alteration of contexts and context conditions, and law-making and legislation).
Criteria of success for political governance (political steering):	In general: e.g., the political governance (steering) supports the "self-rationale(s)" of the higher education system.
	In particular: e.g., the political governance ("public promotion") supports the sciences and arts in higher education by supporting the effectiveness (for example, quality) of university research (R&D).

Source: Authors' own conceptualization and visualization (based on Campbell, 2001, p. 435, and Campbell and Carayannis, 2013, p. 38).

4

Governance of Higher Education, "Epistemic Governance" and the Governance of Quality Enhancement in Higher Education Institutions

As Jürgen Enders and Jon File (2010, p. 15) explain that governance involves the interaction of actors, rules and regulations. It addresses key questions about decision-making processes and the "rules of the game". These questions can be explored at various levels, such as, within universities (internal governance) and at "the macro- or system-level (external governance)." Geuna et al. (2003, p. 393) describe governance as a "source of external influence" on higher education. They state that the governance system for science, comprising institutions both within and outside the state, shapes the incentives, social norms and priorities of scientific research, and significantly influences "scientific research activities and the organization of scientific institutions". Building on Geuna et al. (2003), we can extend their ideas further by suggesting that *the external governance of higher education and institutions can be seen as an "external environment influencing higher education behavior. Internal governance can be viewed as an "internal environment" within institutions"* (Campbell and Carayannis, 2013, p. 40).[34] This approach is

[34] In systems theory terms, "external governance" is part of the "external environment" for institutions, sectors, and systems. These entities must adapt to or consider their external social and societal environments, including efforts at external governance (see Willke, 1989, and Luhmann, 1988).

DOI: 10.1201/9781315209234-4

crucial for understanding academic faculty behavior and the academic profession.[35]

Ferlie et al. (2009) focus on two key approaches to higher education governance, relating them to the evolving roles of the nation-state: "New Public Management" and "Network Governance". These represent the main narratives in public service reform.[36] *In this context, the question arises whether New Public Management in its various forms, has become a governance standard in higher education* (Park, 2013; Campbell and Stransky-Can, 2020). The following compares the key characteristics of New Public Management and Network Governance (see Campbell and Carayannis, 2013, pp. 47-48).[37]

1. Key *characteristics of the "New Public Management" narrative* (Ferlie et al., 2009, pp. 14-15):[38]
   1.1 "Market based reforms", such as competition for attracting students and securing research funding.
   1.2 "Tightening budgetary constraints", which involves focusing on efficiency and ensuring good value for money.
   1.3 "Stress on performance" (for example, implementing advanced methods to measure performance.
   1.4 "Concentrating funds in the top performing higher education institutions".
   1.5 "The Ministry and its agencies guide the system by setting targets and performance contracts".
   1.6 "Higher education governance", which includes empowering strong university leadership and reducing faculty and influence.
   1.7 "Defined managerial roles" emphasizing the importance of clear managerial roles within higher education institutions.

35 "Governance often appears as a flexible and welcome alternative to the traditional mechanisms of hierarchy and market. It is presented as a self-organized process that takes place in inter-organizational networks of autonomous actors from the state, economy, and civil society, who are often implicitly described as equals in political science and sociological literature".
(Biegelbauer, 2010, p. 11 quote translated into English by DFJC). Compare with Krücken 2003a and 2003b, and Krücken et al., 2007.

36 In an earlier article for *Higher Education,* Ferlie et al. (2008, pp. 334-340) also refer to "Three main narratives of public sector reforms and how they apply to higher education". Ferlie et al. initially included a third approach, the "Neo-Weberian narrative", alongside "New Public Management" and "Network Governance". However, they later simplified their framework to focus on just two main narratives (Ferlie et al., 2009, pp. 12-18).

37 See also Ferlie et al., 1996.

38 New Public Management in this context includes numerous references to quality assurance, enhancement, and management (refer to the argumentation that follows in this Chapter) .

1.8 "Expanding performance-based pay for faculty and adopting private-style Human Resource Management".

2. *Crucial characteristics of the "Network Governance" narrative* (Ferlie et al., 2009, p. 17):[39]
   2.1 Establishing and supporting networks between institutions" and between higher education institutions and other social actors.
   2.2 Networks foster and maintain substantial self steering and self organizing capacity.[40]
   2.3 Some networks specifically focus on "joint problem recognition, "joint problem-solving", "organizational learning" and sharing "good practices" or "leading-edge knowledge";
   2.4 "External control systems" align with light touch systems' approaches and rely on professional self-regulation.
   2.5 The governance of higher education systems can also be network-based, with "governmental and non-governmental organizations" interacting at "different levels" and in "different functional areas";
   2.6 The Ministry of Education, equivalent ministries and similar "HE agencies" tend to influence higher education institutions indirectly, adopting a more "'hands off' style of system management at the national level";
   2.7 The "senior management style" emphasizes "softer leadership skills", "distributed leadership", and "team based approaches", contrasting with more "highly individualized management typical" for New Public Management;
   2.8 "Human resources management systems" prioritize rewarding "high performing teams" rather than "individuals".

The "standard model" is to assert that there exist two main narratives of governance in higher education, the New Public Management and Network Governance (Ferlie et al., 2008 and 2009). *The approach here is to add "Epistemic Governance" as a third narrative, or as a third paradigm of governance of higher education* (Campbell and Carayannis, 2013 and 2016b). *This should support a transformation of governance and a governance of transformation* (Burkert et al., 2024; Campbell, 2024) *and enable principles of "systemic governance" in higher education.*

[39] Compared with Rhodes (1996). Some aspects of the "Network Governance" narrative (Ferlie et al., 2009) are directly discussed in Rhodes' (1996) concept of "Governing without Government".

[40] This enables conceptual connections to the ideas of *self-rationale, self-organization, self-steering, and self-governance in systems theory.*

Alice B.M. Vadrot (2011) is the first to have introduced the concept of "Epistemic Governance" into academic discourse, providing an insightful discussion on the topic:

> The concept of "epistemic governance" addresses power dynamics in creating, structuring, and coordinating knowledge on socio-ecological issues. It examines how policy-making and scientific knowledge production define and tackle these problems, considering the influence of power relations on knowledge creation. The effort to develop a shared knowledge base jointly by political actors and scientific communities requires the "epistemic governance" framework. This framework enables the analysis of the re-production of interconnected socio-economic and natural systems. Finally, knowledge production and use are intertwined with relational, structural, and soft power dynamics, as well as the complex, non-linear relationship between science and policy (Alice B.M. Vadrot, 2011, p. 50).[41]

In our examination of governance in higher education, we adopt and refine the concept of epistemic governance. We define it as addressing the foundational epistemic structures, bases, or paradigms of organizations, institutions, or systems requiring governance. These aspects reveal the fundamental rationale or logic guiding their behavior.

Is it possible for any organization, institution, or system to function without an underlying epistemic structure? Some organizations or institutions might lack a clear epistemic structure, it is unlikely for entire systems or sectors. This underlying structure can be explicit or implicit. Additionally, it can be diverse, contradictory, or heterogeneous, indicating multiple coexisting paradigms. These paradigms may compete, evolve together, and create complex patterns of cooperation and competition, resulting in "co-opetition" (Brandenburger and Nalebuff, 1997).

If organizations, institutions, or systems are based on an underlying epistemic structure, sustainable and effective governance, whether internal or external, is only achievable in the long term when this epistemic foundation is recognized. Objectives and policies can only be realistically formulated, developed, and implemented within the governance process if these underlying epistemic structures are considered. If the

[41] See Vadrot (2008) for an earlier writing of Vadrot. Furthermore, compare with Vadrot (2014). Review also Langlet and Vadrot (2024).

epistemic structure is diverse and pluralistic, governance faces additional challenges for governance. *Good, sustainable and effective governance whether external or internal,is achievable in the long term only when the underlying epistemic structure, base, paradigms are addressed"* (Campbell and Carayannis, 2013, p. 27).

Universities, higher education institutions are crucial to advancing knowledge economies, societies, and democracies.Their epistemic structure encompasses knowledge creation and production across various functions, including tertiary education, research (basic, interdisciplinary and transdisciplinary) , and third mission activities like innovation. "Knowledge paradigms" in higher education refer to the fundamental frameworks and principles that drive knowledge creation and dissemination within these sectors. So, what *are the paradigms of knowledge that drive higher education?* (Campbell and Carayannis, 2013, p. 27).

Knowledge creation and production play a fundamental role in higher education institutions and the entire system. Thus, the type of knowledge produced and the underlying paradigms shaping it are crucial components of the epistemic structure of higher education. Effective epistemic governance of higher education necessitates addressing the underlying epistemic structure and the paradigms driving the produced knowledge. This governance is closely tied to the knowledge paradigms of higher education. When aligned with quality management practices, these paradigms can be translated into quality dimensions for assessment and improvement. The one core definition here is: *Epistemic governance in higher education requires addressing the underlying epistemic structure and knowledge paradigms that drive higher education* (Campbell and Carayannis, 2913, p. 27). Refer to Figure E for a graphical visualization.

Just as knowledge paradigms are vital for epistemic governance, quality dimensions play a similar role in quality management. This underscores that any effective quality management approach in higher education must consider its underlying epistemic structure, even indirectly. Without this, defining quality in higher education becomes difficult. Quality management that ignores the epistemic aspect is limited in effectiveness within higher education.Therefore, *comprehensive and sustainable quality management in higher education must consider the underlying epistemic structure at least implicitly* (Campbell and Carayannis, 2013, p. 27).

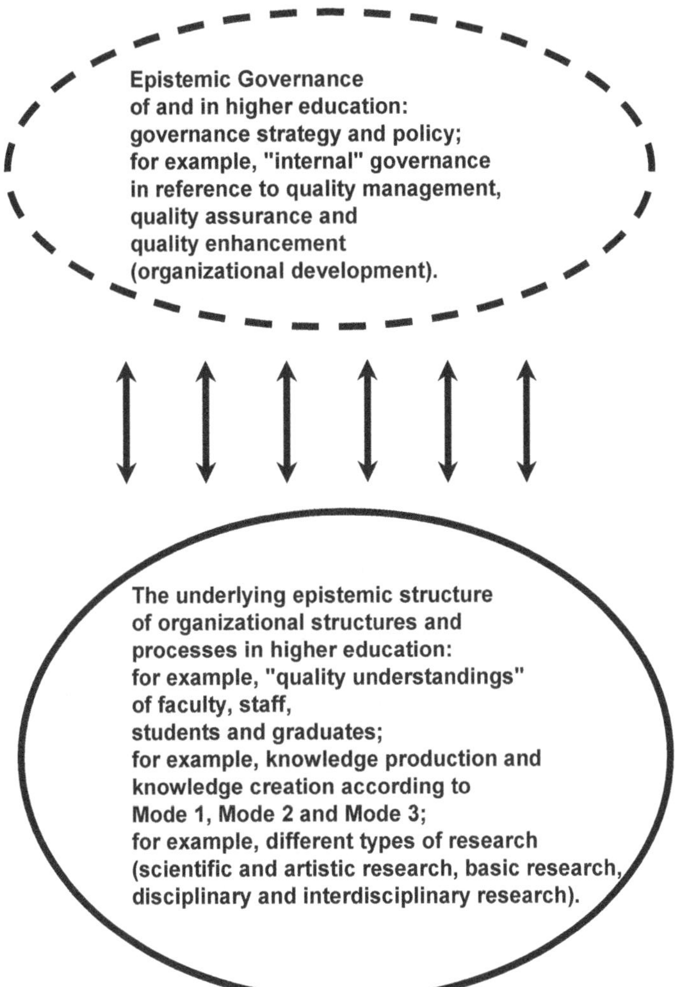

Fig. E Governance, "Epistemic Governance" of and in Higher Education.
Source: Authors' own conceptualization and visualization based on Campbell & Carayannis (2013) and Carayannis & Campbell (2021).

The concept of quality in higher education encompasses various dimensions.(1) Sometimes viewed as a "black box," quality can be approached procedurally, allowing relevant communities to define it according to their understanding. (2) *Efficiency* focuses on maximizing output while minimizing input. (3) *Relevance* encompasses its significance for academic research communities and its practical application in society and the economy. (4) *Viability or sustainability* evaluates the organizational or institutional context within which quality is maintained.

Higher Education Institutions and the higher education systems are crucial drivers for the further evolution of knowledge economy, knowledge society, and knowledge democracy (Campbell, 2019). *It is crucial to define the governance of higher education. A key aspect is linking quality assurance and enhancement with organizational development in higher education.*

The key concept here is "Epistemic Governance" (Campbell and Carayannis, 2013, 2016a and 2016b). Epistemic Governance connects academic quality understandings with internal and external governance measures, creating mutual references. Governance that ignores epistemic aspects cannot lead to long-term sustainable development in academia. Consequently, this implies:

Here is a core definition of Epistemic Governance:

"Epistemic governance in higher education requires addressing the underlying epistemic structure and knowledge paradigms that drive higher education" (Campbell and Carayannis, 2013, p. 27). *Quality management* rests on two principles: *quality assurance* and *quality enhancement*. Quality assurance functions as a necessary condition, focusing on quality control, while quality enhancement is the "sufficient condition" promoting quality development.[42] In the short term, quality control is important, but in the long term, *quality enhancement is crucial* (Carayannis and Campbell, 2021). Balancing *quality development and organizational development, such as in higher education institutions is the goal of Epistemic Governance.* Epistemic Governance is essential for advancing teaching, research and learning at Higher Education Institutions, as well as their innovative performance.Policy moderation is also important (Kuhlmann, 1998). Additionally, the concept of "Epistemic Democracy" exists (Rothstein, 2011; Schwartzberg, 2015). State institutions of a state (government) can be examined for underlying epistemic structures and patterns.

Quality management (QM), or a quality management system (QM system), integrates quality assurance and enhancement. In higher education, it combines elements of accreditation, evaluation, audit, and certification. Quality management involves the internal governance or steering of higher education, and it can also refer to the external governance by other actors, such as the state (Campbell and Carayannis, 2013, p. 55).

---

[42] See also Burkert et al., 2016.

Quality management, encompassing both quality assurance and enhancement, should be rooted in quality. One approach to *quality is by identifying specific "quality dimensions"*, which are often deemed important for higher education (see Figure F):

(1) *"Quality"* is often ambiguous or seen as a "black box" due to competing understandings. One way to define quality is through procedural methods, allowing relevant communities to determine what quality means to them. This approach emphasizes flexibility and a community-driven understanding of quality. This approach emphasizes a flexible, community-driven method for defining and assessing quality. As a result, each *specific definition of quality reflects the unique "quality understandings" of various disciplines and communities, which can be seen as "quality cultures"*.[43] For instance, "quality understandings" in research can vary across basic and applied research; disciplinary, interdisciplinary and transdisciplinary research as well as scientific and artistic research (Carayannis and Campbell, 2015). *This diversity of "quality understandings" justifies the need for varied governance approaches within and for higher education. The diversity of quality understandings is connected to a diversity of governance in higher education* (Campbell, 2024; Campbell et al., 2021).[44]

(2) *"Efficiency"* means achieving the highest possible output with the least amount of input, essentially optimizing the input-output ratio.

(3) *"Relevance"* can be understood in two ways. its significance is within higher education, like its importance to academic research communities and its applicability in real-world contexts, including its relevance to society and the economy.

(4) *"Viability"*, often called "sustainability", assesses the frameworks supporting long-term sustainability in higher education. It considers the structures and processes that facilitate education, research, and societal engagement, ensuring institutions can fulfill their missions while adapting to changing needs.

---

[43] On the connection between Epistemic Governance, quality understandings, and quality cultures:

"When epistemic governance seeks to grasp the self-understanding of institutions and organizations, especially in relation to knowledge production (research), another conceptual step is appropriate: the 'quality culture' of a university (higher education institution) is also an integral part of its self-understanding" (Campbell, 2017, p. 159).

[44] The University of Applied Arts Vienna (in Austria) set up a unit for quality enhancement, named "Universitätsentwicklung und Qualitätskultur", *University Development and Quality Culture* (https://www.dieangewandte.at/uqe).

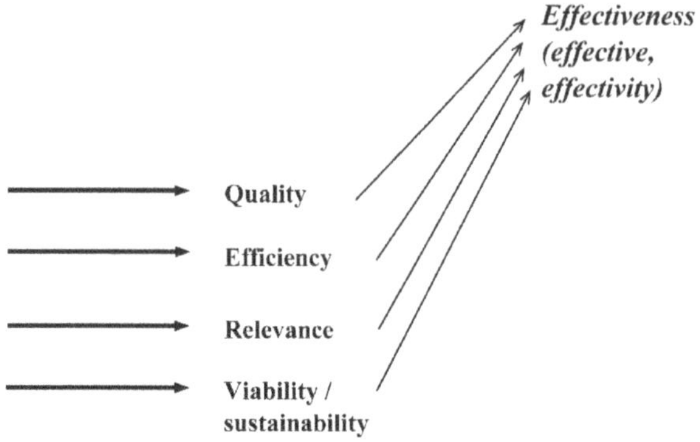

Fig. F Spectrum of Different "Quality Dimensions" for Governance of and in Higher Education.
Source: Authors' own conceptualization and visualizationbased on Campbell & Carayannis (2013) and Carayannis & Campbell (2021).

(5) *"Effectiveness"* measures how well objectives or goals are achieved. Unlike efficiency, it is more complex to quantify and can be assessed based on factors like quality, efficiency, relevance, and sustainability. This results in varying effectiveness profiles depending on the context and priorities. "Effectiveness, as an advanced dimension, assesses how well effective university research meets its objectives. Often used in policy terms, it expresses the degree of achievement of specific research goals. Consistently applying the concept of effectiveness can be challenging. We argue that effectiveness can be modeled by combining 'first-level' dimensions, resulting in effectiveness profiles for different institutions or disciplines. Consequently, some institutions excel in efficiency, while others may stand out in relevance" (Campbell, 2003, p. 130).

Quality management in higher education includes both internal and external governance. Internally, it ensures high standards in teaching, research, and activities within institutions. Externally, it involves oversight by entities like government bodies or accrediting agencies to meet quality standards and objectives. Quality management in practice encompasses various well-established approaches (OECD, 2008, pp. 265-277; Teichler, 2006), including *certification accreditation, assessment or evaluation, and audit or review* (see Figure G for a visualization).

Stefanie Ernst (2008, p. 27) proposed a difference between "evaluation" and "quality management". Evaluation primarily aims at improving processes, whereas quality management is inclined towards stabilizing these.

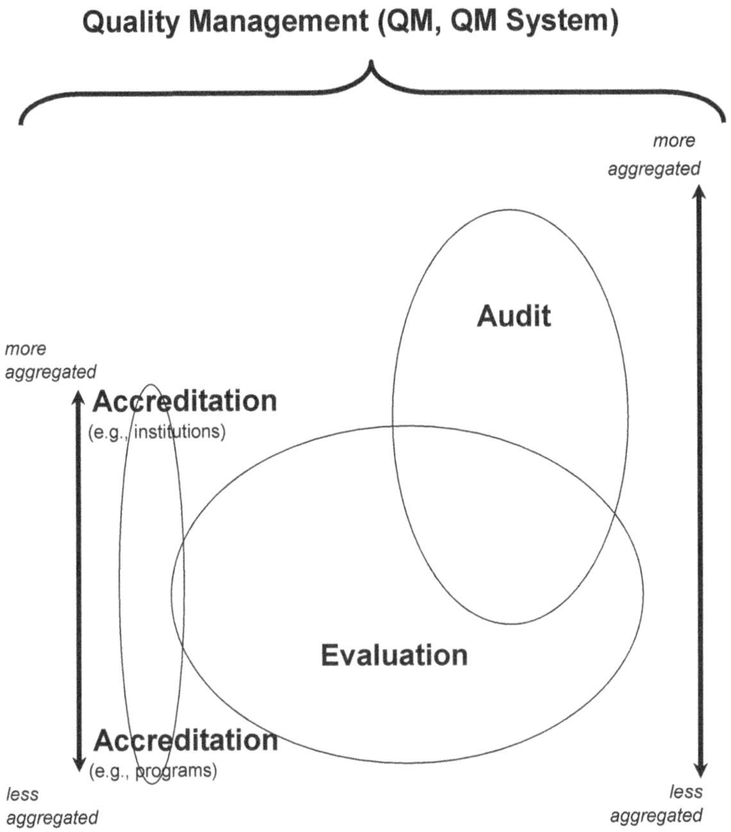

Fig. G Quality Management Governance (QM, QM Systems) for and in Higher Education Institutions (HEIs).

Source: Authors' own conceptualization and visualization based on Campbell & Carayannis (2013) and Carayannis & Campbell (2021).

In practice, there often is an overlap between the various approaches and procedures used for quality assurance and enhancement in higher education. Conceptually, it is often challenging to clearly define boundaries between these approaches. *When there is a conceptual overlap, it may lead to the recognition of a "two-tier" or "two-level" system of quality assurance and enhancement.* For instance, an institutional audit may examine whether the institution has processes in place for self-evaluation of teaching and research. This could be seen as a form of *"meta-evaluation"* (*evaluation of evaluations,* perhaps also an *evaluation of self-evaluations*), where the audit provides a broader or different perspective on the evaluations conducted by the institution itself.

Epistemic Governance has the capability and capacity to ensure that all these specific measures of quality assurance and quality enhancement are interconnected with the underlying quality understandings (and quality dimensions) in higher education. Epistemic Governance shows a route into a future of higher education governance.

5

Innovation: Linear and Non-Linear Models of Innovation and Innovation Networks of Entrepreneurial Universities and Academic Firms

Research (R&D, research and experimental development) represents a form of knowledge creation or knowledge production. Innovation can be described in different ways (Chux and Gebhardt, 2021; Ibanez et al., 2024; Ouragini et al., 2024). However, innovation is often *associated with knowledge application.* Consequently, an *innovation process* can be described as: "Innovation = is a process, where knowledge is being used for the purpose of a (new) application and where also (new) knowledge is being created" (Campbell and Pantelić, 2020a, p. 3). Consistently, then a *learning process* can be seen as: "Learning = is a process, where knowledge is being used for the purpose of a (new) application and where also (new) knowledge is being created, and where an improvement, betterment, advancement, or a reform are being (or were) achieved" (Campbell and Pantelić, 2020a, p. 3).

Within the domain of innovation there exist various models of innovation, including linear and non-linear ones such as the Triple Helix (Etzkowitz and Leydesdorff, 2000), Quadruple Helix (Carayannis and Campbell, 2009), and Quintuple Helix (Carayannis and Campbell, 2010) models, along with Creative Knowledge Environments (Hemlin et al., 2004).[45] The linear model of innovation is often being attributed to Vannevar Bush (1945)[46], although he never explicitly used the term "innovation" in his report "Science: The Endless Frontier" from 1945.

[45] For the development of a comparative framework of analysis across these different Helical models, see Cai and Lattu (2021).

[46] For example, this was asserted by Narin et al. (1997, p. 318).

DOI: 10.1201/9781315209234-5

Despite this, it is being argued that Bush may have implicitly referred to innovation. *The concept of linear innovation involves a sequential process where scientific research (scientific research in combination with artistic research) possibly leads to technological innovation, which then leads to economic growth.*

The *linear model of innovation* has the following expression: In terms of research, universities and the higher education sector predominantly focus on basic research, typically funded publicly. This basic research gradually disseminates from the higher education sector into society and the economy, often in a laissez-faire manner. Eventually, businesses adopt some of these research findings and transform them into applied research and experimental development, aiming to create profitable commercial products and services. Applied research and experimental development are mainly conducted within the business sector and are typically privately financed, although public funding may be more significant in less developed industries and economies. Broader conceptualizations of research therefore also use the acronym R&D, meaning "research" (basic research and applied research) and "experimental development" (OECD, 1989 and 1994).

In contrast to the linear approach, *non-linear models of innovation* emphasize simultaneous engagement in basic research, applied research, and experimental development by both universities (higher education institutions) and firms. This means that instead of a strict first–then sequence, there is a recognition of parallel effects where these activities occur concurrently.

"In contrast to the linear model, the paralleling of basic research, applied research and experimental development demands that the different R&D activities should be considered, to phrase it in a challenging language, as 'parallel processes'" (Campbell and Güttel, 2005, p. 167; see also Campbell, 2000).

Within the framework of *"non-linear innovation"* models the focus is on the following processes: While distinct "linear innovation lines" may still function at the organizational or institutional micro-level (meso-level); at the meso-level or macro-level, there are opportunities for participation in various linear innovation lines at different stages. This means that universities and firms engage in basic research, applied research, and experimental development (R&D) simultaneously, resulting in a *"parallelization"* of research and development and innovation. Consequently, *the traditional sequential first–then relationship transforms into a "first-first" relationship,* where multiple stages of innovation occur concurrently. *"As a result, universities and firms conduct basic research, applied research and experimental development simultaneously. The sequential first-then relationship transforms into a parallel 'first-first' relationship"* (Campbell and Carayannis, 2013, p. 30). See Figure H for a visualization.

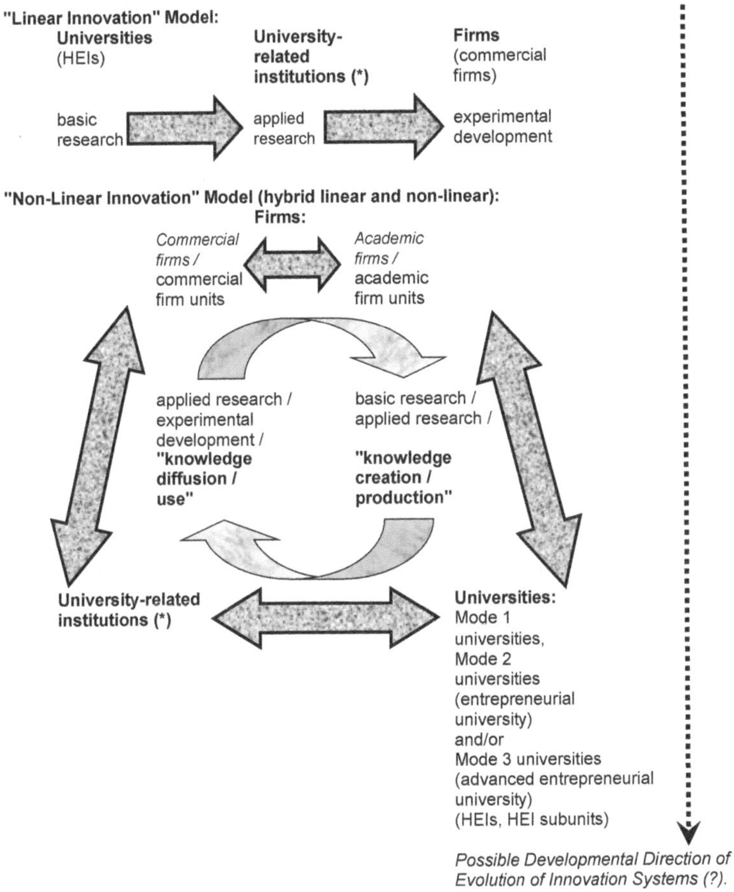

(*) University-related can be translated into German as "außeruniversitär" (Campbell, 2003, p. 99).

Fig. H Evolution of "Linear Innovations Systems" to a Combination, Recombination and Hybrid Mix of "Linear and Non-Linear Innovation Systems".

Source: Source: Authors' own conceptualization and visualization based on Campbell & Carayannis (2013) and Carayannis & Campbell (2021)

"Here, we may witness an overlap of linear and non-linear innovation, creating an overall system of non-linear innovation." Campbell and Carayannis, 2013, p. 30 (see again Figure H[47]; compare with Carayannis and Campbell, 2012, p. 25).

In addition, should the concept of "Technology Life Cycles" (Tassey, 2000, 2013 and 2018) be analytically employed, then another non-linear aspect may be added: *In a non-linear innovation context, organizations or institutions (including firms) operate simultaneously across different technology life cycles at different maturity levels or maturity stages.*

A crucial challenge lies in designing research frameworks that foster cross-learning and cross-fertilization between different linear innovation lines or research lines. This blending of linear and non-linear innovation creates a "comprehensive" system of non-linear innovation, *leading to a more complex innovation landscape of innovation ecosystems*. For instance, when universities delve into applied research and firms into basic research, *it necessitates more hybrid and network-based connections between universities, firms, and other entities such as university-related institutions, media-based and culture-based public, and civil society, or the whole knowledge democracy*. The Quintuple Helix concept further integrates by incorporating the environment into the innovation system, embedding the Quadruple Helix and Triple Helix within an environment context.

"When universities conduct applied research and firms engage in basic research, it fosters opportunities and needs for hybrid network-based linkages.This includes connections between universities, firms, university-related institutions, and the "media-based and culture-based public" and "civil society" within Quadruple Helix innovation arrangements (Carayannis and Campbell, 2009, p. 207; 2011, pp. 13-14; 2012, pp. 13-14).[48] The Quintuple Helix, ultimately, integrates the 'environment or the natural environments' into the overall architecture of innovation systems (Carayannis and Campbell, 2010, pp. 61-62). 'The Quintuple Helix finally embeds the Quadruple Helix (and the Triple Helix) in context of the environment or the natural environments' (Carayannis and Campbell, 2010, pp. 61-62)" (Campbell and Carayannis, 2013, p. 30; 2019).[49]

---

[47] In Figure H, so-called "university-related institutions" are listed. In German they would be called "außeruniversitär". This indicates academic-style organizations that do not belong to the higher education sector since they do not perform teaching activities.

[48] See also: Danilda et al., 2009.

[49] "The Quintuple Helix model is interdisciplinary and transdisciplinary at the same time: the complexity of the five-helix structure implies that a full analytical understanding of all helices requires the continuous involvement of the whole disciplinary spectrum, ranging

The Triple Helix model examines the relationships among universities, industry, and government in knowledge production and innovation. It emphasizes the importance of "trilateral networks and hybrid organizations" to facilitate these interactions and collaborations (Etzkowitz and Leydesdorff, 2000, p. 111). Universities are encouraged to develop business management skills, while firms are urged to engage more with the academic world, leading to the emergence of concepts like the "Entrepreneurial University" (Etzkowitz, 2003) and the "Academic Firm" (Campbell and Güttel, 2005, pp. 170-172; Campbell and Carayannis, 2016a and 2023a). This dynamic fosters niches and opportunities for collaboration and innovation between academia and industry.

Academic firms and commercial firms can exist alongside each other and evolve together. *While commercial firms prioritize profit and profit maximization, academic firms are more concerned with creating social environments conducive to academic-style knowledge creation and creative knowledge production, similar to university settings*. Academic firms may also allocate some of their efforts to academic-style-research activities and academic publication formats, further aligning with the academic context. The concept of the academic firm focuses on creating social environments that foster academic (academic-style) knowledge creation and creative knowledge production similar to university settings (Campbell and Carayannis, 2013, p. 31; compare with Carayannis and Campbell. 2009, pp. 211-212).

An academic firm can take various forms: it could be an entire organization, a division within a commercial firm, or exhibit specific characteristics commonly associated with academic environments. These characteristics may include provisions for *continuing education, lifelong learning opportunities,* and facilitating employee absences for activities like sabbaticals. Additionally, academic firms may allow for split "cross-employment", where employees work simultaneously for multiple organizations, particularly academic institutions such as higher education institutions.

One basic idea of the academic firm is to create innovations by conducting and engaging in academic-style research processes. This further defines an interest for the academic firm to participate in wider networks, composed of organizations, such as higher education

---

from the natural sciences (because of the natural environment) to the social sciences and humanities (because of society, democracy and the economy). The Quintuple Helix also is transdisciplinary, since it can be used as a frame of reference for decision-making in connection to knowledge, innovation and the (natural) environment" (Carayannis and Campbell, 2011, p. 62). Refer to Campbell and Campbell (2011, pp. 15-16, 23-27; 2017; 2019).

institutions and other types of organizations, such as "university-related institutions" or civil-society-organizations.

"Creative Knowledge Environments" (CKEs) are defined as environments, contexts and surroundings that positively influence individuals engaged in creative work, aiming to produce new knowledge or innovations, whether working alone or in teams, within a single organization or in collaboration with others (Hemlin et al., 2004, p. 1).

Building on the concept of "creative knowledge environments", *an academic firm develops "internal-creative-knowledge-environments" that mimick academic settings. This fosters creative research, innovation and other forms of creative knowledge production.*

Universities, especially entrepreneurial ones, and academic firms, should collaborate closely without merging completely.. This collaboration fosters diverse and pluralistic *networks* for knowledge and innovation, facilitating development and creation. Within these "innovation networks," entrepreneurial universities and academic firms act as "knowledge clusters" (Carayannis and Campbell, 2006b).

5.1 The Mode 1 and Mode 2 in Knowledge Production

Gibbons et al. (1994) introduced Mode 1 and Mode 2 knowledge production. *Mode 1 refers to basic university research with a disciplinary focus, conducted without prioritizing innovation or practical application.* This research is typically long-term and confined within disciplinary boundaries (Gibbons et al., 1994, pp. 1, 3, 8, 24, 33-34, 43-44, 167).

Mode 1, the traditional knowledge production model in universities is being challenged by *Mode 2. Mode 2 emphasizes applying knowledge, interdisciplinary collaboration, diverse organizational structures, social responsibility, and quality assurance.* Mode 2, which evolved from Mode 1 integrating university research with societal and economic needs. It emphasizes problem-solving and practical applications, reframing basic research, applied research, and experimental development within real-world contexts.

Mode 2 of knowledge production is driven by principles like: (1) application-contextualized knowledge; (2) transdisciplinarity; (3) heterogeneity and organizational diversity; (4) social accountability and reflexivity; and (5) quality control (Gibbons et al., 1994, pp. 3-8, 167). Mode 2 grew out of Mode 1, and Mode 2 emerged from and co-evolves with Mode 1 (Gibbons et al., 1994, pp. 14, 17) (Campbell and Carayannis, 2013, p. 32).

There is a growing overlap between *the processes of "discovering" new knowledge and "applying" existing knowledge, including experimentation and simulation and "knowledge design".* Applying knowledge can lead to the discovery of new knowledge, blurring the lines between traditional linear innovation and nonlinear approaches. This overlapping suggests that applying knowledge is essential for advancing basic research, as it contributes to the development and refinement of theories. Moreover, it supports continuous innovation in both Mode 1 and Mode 2 knowledge production. Quality remains crucial for both modes of knowledge production (on Mode 1 and Mode 2, see furthermore: Nowotny et al., 2001, 2003 and 2006; Scott, 2009; Campbell, 2006b, pp. 71-73, 91-92; Carayannis and Campbell, 2010, pp. 48-52).

In Mode 1, quality is defined as academic excellence, which entails providing a comprehensive understanding of the world based on fundamental principles, evaluated by academic communities using disciplinary peer review systems. Conversely, in Mode 2, quality is defined by problem-solving, emphasizing the usefulness and effectiveness of solutions for society, economy and democracy, judged by both knowledge producers and knowledge users including communities of knowledge producers and users. These varying definitions of quality reflect distinct knowledge paradigms in higher education, similar to those discussed by Kuhn (1962).

> "Quality in Mode 1 is defined as academic excellence, offering a comprehensive explanation of the world and society based on basic principles, judged by academic communities through a disciplinary peer review system. In Mode 2 quality is defined by problem-solving, that is useful, efficient, and effective for society. It is judged by both knowledge producers and users. Mode 1 and Mode 2 can be seen as 'knowledge paradigms' underpinning higher education (see also Kuhn, 1962)" (Campbell and Carayannis, 2013, p. 32).

5.2 Mode 3 Knowledge Production in Mode 3 Universities (Higher Education Institutions)

"Mode 3" introduces the concept of diverse knowledge types and paradigms coexisting and evolving together. It recognizes *the co-existence, co-learning and co-creation of various knowledge and innovation modes, emphasizing the importance of this diversity.* This paradigm highlights the significance of applied knowledge in innovation and underscores the need for different knowledge paradigms to evolve harmoniously.

" ...'Mode 3' emphasizes the co-existence and co-evolution of different knowledge and innovation paradigms. A key hypothesis is that the *competitiveness of a knowledge system is determined by its ability to combine and integrate various knowledge and innovation modes through co-evolution, co-specialization and co-opetition dynamics* (e.g. Mode 1, Mode 2, Triple Helix, linear and non-linear innovation)" (Carayannis and Campbell, 2009, p. 223) (Campbell and Carayannis, 2013, p. 33).

Mode 3's diverse structure aligns with with democratic principles, suggesting that multiple knowledge modes coexist within a "Democracy of Knowledge". This interrelation emphasizes the importance of pluralism in knowledge modes (Carayannis and Campbell, 2009, pp. 208, 224). *There is an anticipated interlinkage between "diversity of knowledge" and "political pluralism in a democracy"*. This indicates that advanced Mode 3 knowledge and knowledge-based democracies are interconnected. There *is a concept of coevolution between knowledge societies, economies, and democracies*, suggesting they mutually influence and shape one another.

A Mode 3 higher education system embraces and encourages various knowledge and innovation paradigms simultaneously. It values the co-existence, co-development, and evolution of diverse knowledge approaches, essential for progress in education, society, economy, and democracy in the modern knowledge-focused era. Knowledge governance in this system aims to support diverse higher education paradigms both internally and externally. "Mode 3 asserts a congruence between the structures and processes of advanced knowledge and democracy" (Carayannis and Campbell, 2010, p. 52).

> *As a Mode 3 higher education system operates simultaneously according to various paradigms and types of knowledge and innovation. A Mode 3 higher education system views the pluralism, co-existence and co-evolution of various knowledge and innovation paradigms also as positive and as necessary for advancing higher education, society and the economy and democracy in the "age of knowledge'. Epistemic governance addresses various underlying knowledge paradigms within higher education, both externally and internally* (Campbell and Carayannis, 2016b).

In Mode 3 higher education, both Mode 1 and Mode 2 knowledge production methods co-exist and can be innovatively combined in organizational structures like networks. This integration creates sustainable surplus and synergies in knowledge creation, crucial for knowledge societies, economies and the "creativity economy" (Dubina et al., 2012).

All modern higher education systems operate as Mode 3 systems. An "advanced Mode 3 system" explicitly values pluralism, co-existence, and co-evolution of knowledge paradigms. *This approach encourages co-learning between different knowledge and innovation modes,* actively facilitating diverse combinations in knowledge production.

Higher education institutions can adopt *Mode 1, Mode 2 (entrepreneurial university),* or *Mode 3 (advanced entrepreneurial university).* This means integrating both Mode 1 and Mode 2 while fostering creative Mode 3 designs to enhance high-quality knowledge production. These modes can be implemented at the university level or within faculties or departments, offering universities various strategies, options, or development profiles.

> "*Higher education institutions can adopt Mode 1, Mode 2 (entrepreneurial university) or Mode 3 structures, aiming to integrate both Mode 1 and Mode 2. Mode 3 designs foster cross-referencing to enhance high-quality knowledge production.* For example, Mode 1, Mode 2 and Mode 3 can coexist at the university level or within specific sub-levels like faculties or departments. This organizational flexibility offers a wide range of strategies, options and profiles for universities and their sub-units" (Campbell and Carayannis, 2013, p. 34).

There is potential for Mode 1, Mode 2 (entrepreneurial), and Mode 3 universities, along with their sub-units, to coexist and evolve together. This dynamic overlap enables a hybrid integration of linear innovation (Mode 1), entrepreneurial approaches (Mode 2), and non-linear innovation (Mode 3), fostering "Creative Knowledge Environments" (Hemlin et al., 2004).

Creativity is crucial for generating new knowledge in higher education. New knowledge can be seen as creative knowledge, and without creativity, the knowledge needed for the innovation process may face serious limitations (Carayannis and Campbell, 2010, pp. 47-48).

Networks can connect and integrate) Mode 1, Mode 2, and Mode 3 knowledge production.Nationally, universities and their sub-units operating in different modes, can co-evolve with commercial and academic firms, advancing knowledge societies, economies, and knowledge democracies. This dynamic can extend to sub-national, supranational, and transnational levels within multi-level innovation systems (Carayannis and Campbell, 2012, pp. 32-35).

The "Mode 3 University" concept includes epistemic and organizational approaches to combine and co-evolve Mode 1 and Mode 2 knowledge production. This is essential for creating a surplus in knowledge production. Unlike Mode 1, which separates basic and applied research, Mode 3 emphasizes "basic research in the context of application" (Campbell and Carayannis, 2013, pp. 34-35). Mode 3 promotes interdisciplinarity and transdisciplinarity, viewing transdisciplinarity as the practical application of interdisciplinarity. Focusing on basic research allows for further development of these approaches.There are opportunities for interdisciplinary and transdisciplinary networking *between scientific and artistic research (Carayannis and Campbell, 2015) there appear to be further opportunities of an interdisciplinary and transdisciplinary networking for new knowledge creation and innovation.*

> The Mode 3 focuses on *"basic research in the context of application" and promotes* interdisciplinarity[50] and transdisciplinarity.Transdisciplinarity can be seen as the application of interdisciplinarity. This approach highlights a route for further development of transdisciplinarity and interdisciplinarity (Campbell and Carayannis, 2013, pp. 34-35).

Mode 3 universities impact both internal and external governance of higher education. *The co-development of diverse knowledge and innovation modes necessitates a "pluralism of governance" in higher education* (Campbell, 2024; Campbell et al., 2021). To address various knowledge and innovation modes, higher education governance should implement Epistemic Governance.This pluralistic approach provides *a valid governance design for the higher education system.*

[50] Pointing toward a further discussion of interdisciplinarity ("*Interdisziplinarität*"), see also more specifically Markus Arnold (2009, pp. 65-97).

6

New Academic Careers

In Mode 1 universities or their sub-units, academic career paths are well-defined. A distinct "core faculty" group seeks tenure in permanent positions and holds top hierarchical positions (Brechelmacher et al., 2015). Those outside this core often aspire to join it to stay within academia.

In Mode 1, university research prioritizes *"academic excellence", which aligns with* peer review expectations within academic disciplines. This *excellence is closely tied to the outcomes of disciplinary-based peer review.* Mode 1's *linear innovation approach* of Mode 1 implies individuals work either in academia or outside it, such as for firms or societal organizations. Research and development (R&D) and other knowledge production *between universities and firms are often sequential with university work preceding firm-based work.*

Mode 1 academic researchers typically start at a university, either staying there (Geppert et al., 2024; Janger et al., 2019) or moving to a firm, with the possibility of returning to academia later. In Mode 2, the situation is more complex. *Quality focuses on efficient and effective problem-solving, assessed by both knowledge producers and users.* This significantly broadens the spectrum of potential peers. In Mode 2 disciplinary-based peer review loses its primary gate-keeping role, making it *increasingly challenging to define quality criteria and select peers based on quality.*

"While Mode 2 knowledge is equally important as Mode 1, higher education often sees core faculty dominated by the Mode 1 paradigm, pushing Mode 2 paradigms to non-core faculty contexts" (Campbell and Carayannis, 2023, p. 35). While both Mode 1 and Mode 2 knowledge are valuable in higher education, Mode 1 often dominates core faculty, with Mode 2 knowledge relegated to non-core faculty roles. This mismatch leads to Mode 2 being somewhat marginalized. However, for innovation, higher education must comprehensively integrate both Mode 1 and Mode 2 knowledge production.

The Mode 3 knowledge paradigm urges higher education institutions to: (1) Develop a distinct Mode 1 or Mode 2 profile. (2) Consider

DOI: 10.1201/9781315209234-6

combining both modes. (3) Explore creative integration methods. Unlike Modes 1 and 2, *Mode 3 challenges universities to overcome potential deadlocks, opening new pathways for improving quality and advancing development.* Imagine a Mode 3 university as a "mosaic picture" in an artistic installation, with different "tiles" (organizational units) colored in Mode 1, Mode 2, or the "hybrid colors" of Mode 3.[51] Epistemic Governance asserts that when different organizational units understand their guiding principles, this diversity drives creativity. *This diversity of knowledge and innovation modes necessitates a pluralism of governance* (Campbell, 2024).

Implications of a Mode 3 organizational design include: (a) Academic faculty members could integrate elements of both Mode 1 and Mode 2 approaches; (b) Mode 3's non-linear innovation allows academic workers to engage in both basic university research and applied firm research simultaneously, described as *"cross-employment"*[52] *or multi-employment* (Campbell, 2011 and 2013b).[53]

This suggests *that knowledge producers and R&D workers may hold simultaneous employment with multiple organizations or institutions* (Campbell and Carayannis, 2013, p. 36). Cross-employment can take various forms. It may involve networks spanning different higher education institutions or linking universities with firms and civil society organizations.[54]

Cross-employment enhances creativity in knowledge production, creation, and innovation. Academics with cross-employment engage in diverse knowledge production activities, potentially integrating both Mode 1 and Mode 2 knowledge, skills, and competencies. This generates "meta-competences for complex environments" (Zenk et al., 2024). In a Mode 3 university, a single academic faculty member could hold multiple academic employment contracts or multiple job descriptions simultaneously. Overlapping contracts and job descriptions can make the distinction between core and non-core faculty more flexible, open, equitable, and fluid. *This encourages transforming a higher education institution into an arrangement of internal networks. "In a university, operating under Mode 3, the same academic faculty member could hold*

---

[51] Could such a "mosaic picture" of a Mode 3 University also be translated into a *"Picture of Fractals"*?

[52] In this context, compare also with the concept of "cross-retirement" (Blasche and Campbell, 2013).

[53] Again using artistic language: May this create "social amphibia"?

[54] Civil society defines one of the references for the Quadruple Helix innovation system, therefore also co-constituting the Quadruple Helix innovation system concept (Carayannis and Campbell, 2009, p. 207; 2011; 2012, pp. 13-14).

multiple, interplaying academic employment contracts simultaneously" (Campbell and Carayannis, 2013, p. 36).

Cross-employment allows academic faculty to engage in parallel careers both inside and outside higher education, facilitating and sustaining non-linear innovation. This significantly enhances the dynamics of self-organizing, cross-overlapping networks. (Campbell and Carayannis, 2013, p. 36). Cross-employment enables academic faculty and professionals to pursue parallel "cross-careers" simultaneously within and outside of higher education. This means that the same individual can have career paths at two different universities or at a university and a non-academic organization at the same time. *Cross-careers and cross-employment foster hybrid networks and enhance networking and inter-networking capabilities.* By facilitating non-linear innovation, cross-employment contributes significantly to the dynamics of self-organizing and cross-overlapping networks (see Figure I). This implies the need for understanding innovation as a "process of learning" (Campbell and Pantelić, 2020a). It also requires the implementation of "organizational innovations". Cross-employment already represents such an illustration of an organizational innovation. Other examples may be "joint applications" (Campbell and Pantelić, 2020b) or the set-up of "Third Spaces". This benefits the formation of creative "Creative Knowledge Environments" (Hemlin et al., 2004). All of this adds to the whole complex of Third Mission (Compagnuccia and Spigarelli, 2020) and Third Mission governance. Learning should encompass all organizational and systemic structures and processes including "learning in politics" (Biegelbauer, 2013); "policy paradigms and social learning" (Hall, 1993); "institutional learning" (Carayannis et al, 2012b); and "governance learning" (Campbell et al., 2021).

In higher education,this leads to the formation of "Third Spaces" (Smith et al., 2021; Veles et al., 2028). A Third Space involves units *combining individuals with diverse skills and competencies.* For example, university administration may also recruit part-time researchers and teachers to understand university research and teaching experiences and trends.

Cross-employment and cross-careers associated with Modes 1, 2, and 3 are crucial for epistemic governance. The hypothesis is that cross-employment is equally important as the traditional tenure-track model for academic careers. It should be considered a viable, equal option for academic careers within and outside higher education, warranting further discussion. Cross-employment should furthermore be integrated into teaching and curricula. Higher education institutions must teach skills and competences (Ehlers, 2020a and 2020b) such as:

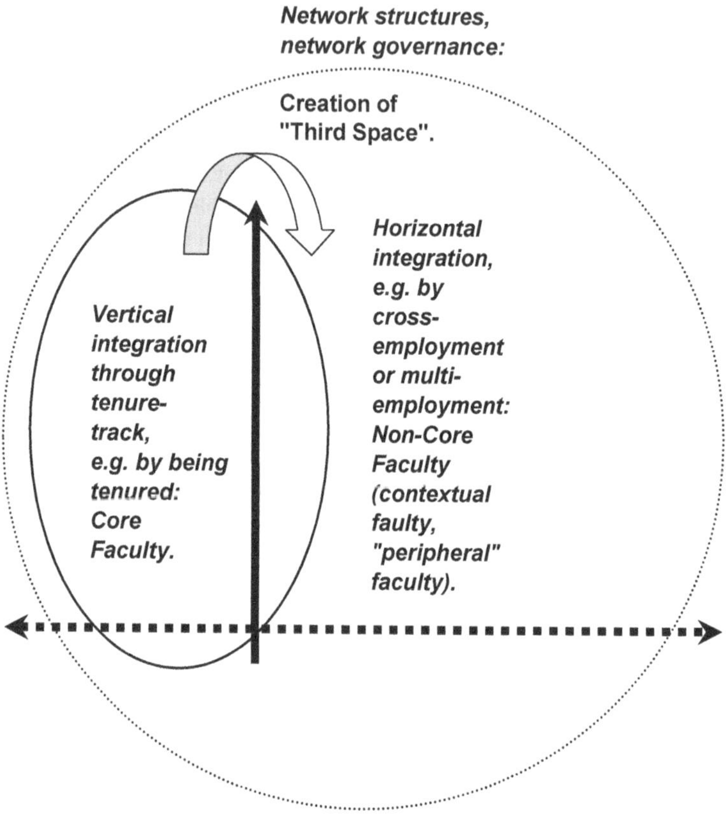

Fig. 1 Vertical Integration through Tenure-Track (Being Tenured), Horizontal Integration through Cross-Employment (Multi-Employment).

Source: Authors' own conceptualization and visualization based on Campbell & Carayannis (2013) and Carayannis & Campbell (2021).

Entrepreneurship education (Compagnuccia and Spigarelli, 2020, p. 20); Meta competences (Zenk et al., 2024); Dual education (Campbell et al., 2023d); Interdisciplinarity and transdisciplinarity; Training for Third Mission and Third Space professionals.

7

The Academic Firm and Innovation

Democracy facilitates innovation (Campbell, 2019), aligning with the concepts of "Knowledge Democracy" (Veld, 2010) and the "Knowledge State" (Campbell, 2006a). The "Quadruple and Quintuple Helix innovation systems" theory underscores the importance of societal, democratic, and ecological contexts for knowledge production, and innovation (Carayannis and Campbell, 2009, 2010, 2021, and 2022; see also Carayannis et al., 2012, 2021, 2022b; and Carayannis and Morawska-Jancelewicz, 2022).

"The 'Academic Firm' is a new type of knowledge-focused, innovation-driven firm operating within Cyber-D4 environments. Theoretical implications of the Quadruple and Quintuple Helix innovation systems are valid for these firms. They *contribute significantly to the evolution of knowledge and innovation, supporting the promotion of knowledge society (Society 5.0), knowledge economy (Industry 5.0) and knowledge democracy (Democracy 5.0)"* (Campbell and Carayannis, 2023a, pp. 300-301).

The "Academic Firm" itself should be seen as an organizational and social innovation in the business world.

The "Academic Firm" prioritizes advancing knowledge production and application, encompassing research, development, and innovation (see Campbell and Güttel, 2005; Campbell et al., 2013; Campbell and Carayannis, 2016a and 2023a). While profit generation is a goal, the focus is on maintaining a balance with knowledge production and adhering to principles of sustainability. Long-term sustainable development guides the strategic direction of the academic firm. In contrast, the "commercial firm" is driven primarily by the goal of maximizing profits. "In contrast, a 'commercial firm', is primarily driven by the goal of maximizing profit" (Campbell and Carayannis, 2023a, p. 301).

"Creative Knowledge Environments" are an established concept (Hemlin et al., 2004). *An academic firm fosters these environments*

DOI: 10.1201/9781315209234-7

internally to enhance innovation and research, simulating academic settings within the firm. By adopting an academic-style atmosphere, the firm encourages and participates in creative and new forms of knowledge production and innovation. This setup fosters creative research and innovation initiatives, allowing employees to engage deeply in research activities similar to those in traditional academic institutions. *Overall, these "internal" creative spaces foster continuous development and forward-thinking approaches in knowledge production and innovation. An academic firm implements and advances "internal-creative-knowledge-environments"resembling academic settings, stimulating creative research, innovation and other forms of knowledge production.*

Between the extremes of the "Academic Firm" and the "Commercial Firm", various combinations of principles are possible. "The Commercial Firm aims to maximize profit, while the Academic Firm focuses on maximizing knowledge and innovation" (Carayannis and Campbell, 2012, p. 27).

Academic firms actively collaborate with universities, higher education institutions, and research organizations. They must be flexible in connecting with both academic and commercial firms. Balancing cooperation and competition, or "co-opetition" (Brandenburger and Nalebuff, 1997), can foster creativity and innovation.

The entrepreneurial (Mode 2) university extends business elements to academia,whereas the academic firm extends academia to the business world. Academic firms focus on knowledge, engage in networks with universities (the higher education sector), promote "academic culture and values", support academic work, and encourage life-long learning and career development through flexible time schemes, and continuing education (Carayannis and Campbell, 2012, p. 27).

The structure and implementation of academic firms provide many organizational possibilities and opportunities (Carayannis and Campbell, 2012, p. 27; 2023a, p. 303):

1 "A whole firm";
2 "A subunit, subdivision, or branch of a 'commercial' firm";
3 "Certain characteristics or elements of a commercial firm".

An entire firm can be structured based on the principles of an academic firm. Alternatively, specific branches of a firm can adopt these principles. Another approach is to focus on certain principles of academic firms and apply them to the entire commercial firm or significant divisions within it. A whole firm may be organized and designed in accordance with the principles of an academic firm. However, it is also possible to organize only subunits (branches) of a firm according to principles of academic firms. Alternatively, the focus may be placed primarily on certain principles of an academic firm, and these principles then can be applied to or across the commercial firm or at least to substantial divisions of the whole (Campbell and Carayannis, 2023a, pp. 303-304).

The term "academic firm" might suggest that it always refers to an entire organization. However, the analysis shows that this would limit the concept. An academic firm can encompass the entire organization or specific units, processes, or principles within it. This flexibility enables hybrid combinations and overlaps between academic firms, focused on knowledge, and commercial firms, focused on profit. "In fact, this event would allow for hybrid combinations and overlapping arrangements between the academic firm (knowledge focused and knowledge driven) and the commercial firm (profit driven)" (Campbell and Carayannis, 2023a, p. 304).

At present, it is challenging to assess the prevalence of academic firms or their principles within contemporary business. The general perception is that commercial firms still dominate organizational structures and development approaches. However, considering the evolution of the economy towards a knowledge-based society, there is a reason to believe that the diffusion and expansion of academic firms are becoming more plausible and potentially beneficial.

This outlines the concept of academic firms, which aim to blend academic values, methods, and research into business practices. *These firms believe that integrating academic research and principles enhances their ability to produce and apply knowledge effectively or to create new knowledge and innovations.* Academic firms view academic research as an integral part of their organization rather than an external activity. They emphasize the connection between research and innovation, promoting a "limited 'scientification' of business R&D" (Campbell and Güttel, 2005, p. 170; see also Carayannis and Campbell, 2009).

When seeking to define the ideal academic firm, certain characteristics and principles emerge. These include integrating academic values and methods into business operations, considering academic research as intrinsic rather than external, emphasizing the link between research and innovation, and promoting a scientific approach to business R&D (refer to Campbell and Carayannis, 2016a, pp. 8-9, and Campbell and Carayannis, 2023a, pp. 304-305):

1. *The academic firm creates internally an "academic ecosystem", which resembles an "internal academic environment", which emulates an "academic world within" a company, to produce similar motivations (as in the context of higher education) for engaging*

knowledge workers in (academic) research and innovation, where the research can then be translated into or connected with innovative knowledge creation and production that provides all the opportunities to be commercially successful.

2. The academic firm *is a unique organization driven by the pursuit of knowledge production and application, particularly through research and innovation.* While profitability is a goal, it is balanced with principles of sustainable and ethical knowledge advancement. These firms operate within a broader knowledge-based ecosystem.
3. The academic firm *is deeply rooted in knowledge, prioritizing its creation and application.* It operates with a focus on non-linear innovation, showing adaptability. Basic research, when applied, is seen as advantageous by the academic firm.
4. The academic firm *integrates academic principles to inspire its employees and foster strong bonds of trust and mutual support between the organization and its staff.* It sees the diverse disciplinary backgrounds and competencies of its employees as advantageous, enhancing their ability to creatively contribute to knowledge production and application. "The academic firm incorporates academic values to motivate its employees and to create bonds of trust and a good relationship between the organization and the individual employees. The academic firm qualifies a disciplinary variety of employee backgrounds including their competences as potential opportunities and assets for creative knowledge production and application" (Campbell and Carayannis, 2023a, p. 305).
5. The academic firm actively seeks partnerships with universities, higher education institutions and other academic research institutions, to access their knowledge resources, such as basic research. *Engaging in diverse networks is valued by the academic firm as a means of both creating knowledge and leveraging opportunities.*
6. The academic firm *supports and promotes academic research (and other academic activities) among its employees,* recognizing, for example, that academic publications can serve as incentives for individuals to document their tacit knowledge. "The academic firm allows and encourages academic research work such as publications as incentives for employees to codify their tacit knowledge" (Campbell and Carayannis, 2023a, p. 305).
7. The academic firm *prioritizes the ongoing education and professional development of its employees, maintaining a positive stance towards flexibility in working hours.* This includes options for full-time, part-time, or even fluctuating schedules, as well as accommodating partial

absences or leaves. Additionally, *the firm explores how continuing education can intersect and enhance career progression and schemes for its employees.* "The academic firm supports continuing education, further education, and lifelong learning of its employees and has in principle a positive attitude in favor of flexibility concerning the load of working hours and their adaptation to the employees and their needs (full-time, part-time, perhaps shifting back-and-forth), and their partial absence or leave. Cross-benefitting cross-connections between careers and career schemes with continuing education are currently being explored by the academic firm" (Campbell and Carayannis, 2023a, p. 305).

8. The academic firm *generally approves and sometimes even encourages its employees to have split employment or "cross-employment" with other academic organizations*, such as universities or other higher education institutions. "The academic firm accepts in principle, in certain situations even promotes, split employment or 'cross-employment' (multi-employment) of its employees with other (academic) organizations or institutions, for example, universities or other higher education institutions" (Campbell and Carayannis, 2023a, p. 305).
9. The academic firm *aims to cultivate "Creative Knowledge Environments" within its internal organization boundaries, fostering spaces where innovation and knowledge creation thrive.* "The academic firm is interested in creating 'Creative Knowledge Environments' (Hemlin et al., 2004) within the internal boundaries of its organization" (Campbell and Carayannis, 2023a, p. 305). Taking these thoughts further, the academic firms promote "internal environments" in terms of internal "creative knowledge environments".

The academic firm holds the potential to revolutionize how knowledge-based economic activities are structured and carried out, ultimately unleashing the creative capabilities of people and individuals. "*The academic firm has the potential of transforming and changing the way knowledge-based and knowledge-oriented economic work is being organized and performed, by unlocking the creative potential of humanity*" (Campbell and Carayannis, 2023a, p. 305).

The concept of the academic firm poses a question: Is it primarily an ideal concept or does it exist in reality? Commercial firms currently define the norm in contemporary business. The empirical validity of the ideas of the academic firm may still need to be confirmed. Academic firms operate in an overall economic environment where success often involves coping with and benefiting from intense competition in an increasingly globalized world.[55]

The idea of "co-opetition" (Brandenburger and Nalebuff, 1997) suggests that success in competition involves creating networks where cooperation and competition overlap. *Between the extremes of the academic firm and the commercial firm, various hybrid organizational forms and combinations are possible, representing a spectrum of approaches.*

"The academic firm represents a challenging proposition for current businesses; the academic firm, however, indicates also routes and paths, for how next-stage changes and future changes and future successes in the world of business and the knowledge economy can be approached and achieved" (Campbell and Carayannis, 2023a, p. 306).

Furthermore: "The academic firm is interested in integrating innovation and entrepreneurship for development, particularly for sustainable development. The academic firm can also be regarded as a support to knowledge democracy" (Campbell and Carayannis, 2023a, p. 306; compare also with Campbell et al., 2015; Carayannis and Campbell, 2014). And: "*In addition, the commercial firm runs the risk of not sufficiently recognizing the requirement and need of an ecological transition that values and supports principles that are environmentally friendly and sensitive"* (Campbell and Carayannis, 2023a, p. 306, with further references to Carayannis and Campbell, 2021, and their approach of speaking of a *"Democracy of Climate"*). The terms of Industry 5.0, Society 5.0 and Democracy 5.0 already exist (Campbell and Carayannis, 2023b). *In this line of thinking, it is also consistent to propose the concept of "Ecology 5.0"* (see Figure J for displaying the actual frame, into which the academic firms are being embedded) .

As the future of labor and education undergo significant changes, the attractiveness of the academic firm as a working environment for "knowledge workers" becomes increasingly crucial. This suggests a mutually beneficial interaction between the economy, society, and democracy, with academic firms playing a significant role in shaping this dynamic. Furthermore, see the following quote that is taken from Bast et al. (2019), pp. 245-246.

"The transformation of Industry 4.0 will destroy labor, and it will create new labor, so finally there may even be more (new) labor (Bast et al., 2019). This requires, however, to reorganize labor and education in innovative and progressive approaches, so that then the net gain of new

[55] Globalization also applies to universities and the higher education systems. "Virtualization" represents one coping strategy with this (Pfeffer, 2012).

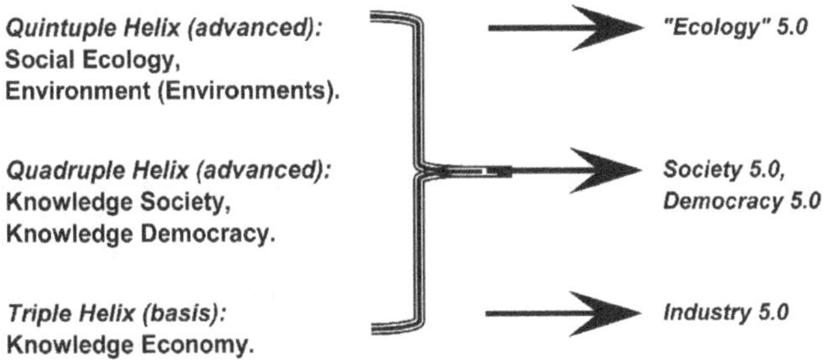

Fig. J The Innovation Architecture of Quadruple and Quintuple Helix Innovation Systems. *Source:* Authors' own conceptualization and visualizationbased on Campbell & Carayannis (2013 and 2023a) and Carayannis & Campbell (2021).

labor has the full potential to outpace the losses of old labor". To prevent labor from being easily replaced by automation or AI, it is essential to continuously develop and enhance human competences. Crucial are multi-facetted competences, combining disciplinary professional knowledge with interdisciplinary and transdisciplinary skills often referred to as "T-competences" and "M-competences" . Creativity and creativity skills are crucial in driving innovation, which is advancing the evolution of knowledge society, economy and democracy. Arts and artistic research represent crucial components in an advanced innovation system. Artificial intelligence will not replace human intelligence, but artificial intelligence will complement human intelligence. The challenge is to organize labor, the economy, society and democracy so that human intelligence leverages artificial intelligence (AI) to support and enhance human intelligence and human labor. The concept is to view AI and human intelligence as co-evolving, where humans in control and making sovereign decisions (termed "Centaur Intelligence"). Artificial intelligence can provide guidance but humans make the final decisions. Advanced knowledge manifests in diverse modes and, requiring political pluralism, a key aspect of democracy. Democracy as an enabler of innovation highlights the connection between knowledge development and the evolution of democracy. Drawing from ancient Greece's democratic polis in Athens, we can speculate that in Industry 4.0, AI and advanced technologies could perform standard tasks, allowing humans to focus on more engaging work.

This can be seen as a Renaissance of engaging work in the Age of Knowledge and Innovation. What are the new and traditional forms of entrepreneurship and creative innovation in Industry 4.0 (or later Industry 5.0)? What does AI-based entrepreneurship mean? Industry 4.0 needs Democracy 5.0. If we consider Art and Democracy, we should also consider the Art of Democracy (Bast et al., 2019, pp. 245-246).

8

The Conclusion: Innovation and Self-Different Fractals

"Fractals" are typically considered to represent "self-similar structures". Two quotes taken from Wikipedia underscore this.[56] [57] This understanding of a fractal, *as something being self-similar,* is also documented in academic literature directly (Carayannis and Campbell, 2011; Graf, 1987). As Hutchinson (1981, p. 1) says: "Sets with non-integral Hausdorff dimensions ... are called fractals by Mandelbrot. Such sets, when they have the additional property of being in some sense either strictly or statistically self-similar, have been used extensively by Mandelbrot and others to model various physical phenomena". In addition to "self-similar" fractals there are also "self-affine fractals" (Kirat and Kocyigit, 2013).

In a fractal-based world, zooming from higher to lower levels never ends, as each level reveals another beneath it. This pattern of endless structural depth continues indefinitely.[58]

[56] "In mathematics, a fractal is a geometric shape containing detailed structure at arbitrarily small scales, usually having a fractal dimension strictly exceeding the topological dimension. *Many fractals appear similar at various scales,* as illustrated in successive magnifications of the Mandelbrot set. *This exhibition of similar patterns at increasingly smaller scales is called self-similarity,* also known as expanding symmetry or unfolding symmetry; if this replication is exactly the same at every scale, as in the Menger sponge, the shape is called affine self-similar" (https://en.wikipedia.org/wiki/Fractal).

[57] "In mathematics, a self-similar object is exactly or approximately similar to a part of itself (i.e., the whole has the same shape as one or more of the parts). Many objects in the real world, such as coastlines, are statistically self-similar: parts of them show the same statistical properties at many scales. *Self-similarity is a typical property of fractals*" (https://en.wikipedia.org/wiki/Self-similarity#:~:text=Self-similarity%20is%20a%20typical%20property%20of%20fractals.%20Scale,the%20object%20that%20is%20similar%20to%20the%20whole).

[58] When such a "fractal-based" or "fractal-structured world" is discussed in reference to "analog" and "digital", what are then the possible (philosophical) implications? Does it mean that a fractal-structured world is more similar to "analog", and it is less similar to "digital"? Perhaps the answer is not so clear, and depends on involved or underlying premises. Therefore, this calls for additional consideration.

DOI: 10.1201/9781315209234-8

Now, the concept being introduced here is that of the so-called "self-different fractals". As far as we know, this term does not really exist as of yet. The reasons for this may be manifold. On the one hand, and radically thought about, a self-different fractal could be considered to represent a contradiction-in-itself.[59] On the other, and more softly formulated, a self-different fractal locates perhaps in an unusual domain of thought or refers perhaps to an uncommon application of ideas in relation to the whole "complex of fractals". *Still the decision was taken to establish explicitly the concept of self-different fractals.* For a simple graphical expression of self-similar and self-different fractals see Figure A (in the Introduction) and Figure K (in the Conclusion). For a complex artistic interpretation of self-similar and self-different fractals, see the graphical images created by Jürgen Pirecki-Giefing in Cluster 1 (Introduction) and Cluster 2 (Conclusion).

We introduce self-different fractals as a philosophical principle, linking directly with the Philosophy of Dialectics. Simplified, the dialectical process moves through these stages: (1) "thesis"; (2) "antithesis"; (3) "synthesis" integrating both.[60] Heinrich Moritz Chalybäus, a German philosopher (1796-1862), is said to have formulated this interaction flow early on. The "Stanford Encyclopedia of Philosophy" defines "Dialectics" as a method of philosophical argument involving a contradictory process between opposing sides.[61] Plato is said to have applied these principles, and Hegel's dialectics also exist in philosophical thought.[62] Even today, the principles of dialectics (thesis, antithesis, synthesis) are applied in modern academic business research (Cunha et al., 2000).

The interplay between "thesis" and "antithesis" may be disruptive, however, if managed or governed properly, such an interaction can act as a driver for creativity, co-creation, innovation, co-development and finally

[59] There is a discussion, whether fractals have to be self-similar or not. One assertion is that is not necessarily the case (https://math.stackexchange.com/questions/4021993/do-fractals-necessarily-have-to-be-self-similar-what-is-the-definition-of-a-fra). There is furthermore a talking about "fractal-like objects" (https://www.sfu.ca/~rpyke/335/W00/21jan.html#:~:text=We%20say%20that%20fractals%20have,for%20the%20fractal%2Dlike%20objects.). Furthermore, fractals often are represented in visualized or artistic approaches (https://www.youtube.com/watch?v=gB9n2gHsHN4).

[60] "This dialectic is sometimes presented in a threefold manner, as first stated by Heinrich Moritz Chalybäus, as comprising three dialectical stages of development: a *thesis*, giving rise to its reaction; an *antithesis*, which contradicts or negates the thesis; and the tension between the two being resolved by means of a *synthesis*" (https://en.wikipedia.org/wiki/Dialectic#:~:text=This%20dialectic%20is%20sometimes%20presented,resolved%20by%20means%20of%20a).

[61] Source: https://plato.stanford.edu/entries/hegel-dialectics/

[62] Georg Wilhelm Friedrich Hegel lived from 1770 until 1831 (https://en.wikipedia.org/wiki/Georg_Wilhelm_Friedrich_Hegel).

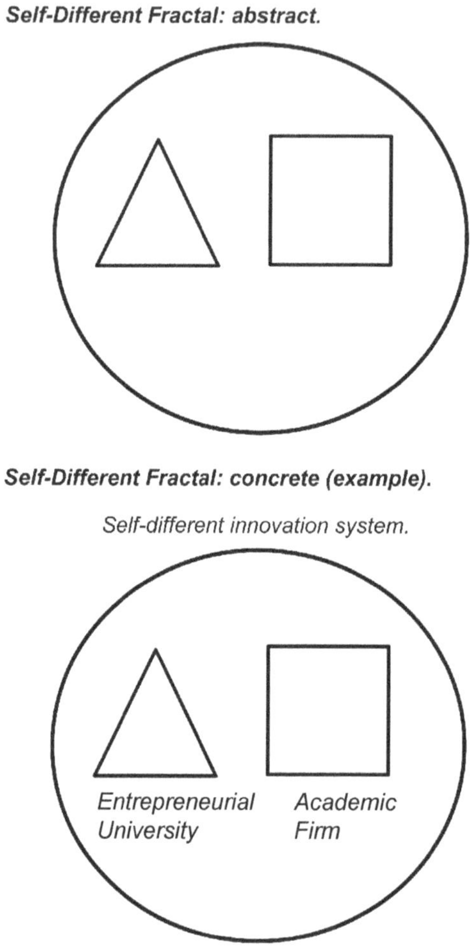

Fig. K Self-Different Fractals Abstract and the Concrete Example of a Self-Different Innovation System.
Source: Authors' own conceptualization and visualization.

development.[63] The assumption is that "self-different" fractal systems, with their co-modality or multi-modality, have enhanced innovation potential. The interaction of different modes may trigger creativity, leading to new forms of knowledge creation, production, and application. *Good governance of a "self-different fractal innovation system" supports*

[63] It is worth recalling that economist Joseph A. Schumpeter linked "creative destruction" to the dynamics of economic cycles (Carayannis and Ziemnowicz, 2007).

the evolution of knowledge, leading to "next"-knowledge. Inspired by a "dialectics of innovation," governance of complex ecosystems of different and co-evolving knowledge and innovation paradigms has the capability to create a "synthesis", where knowledge diversity and innovation diversity interlink in new and productive structures, processes and networks.

This work aims to *translate the traditional "philosophy of dialectics" into the concept of "self-different fractals", linking philosophy with modern innovation theories*. This synthesis connects classical occidental thinking with modern innovation, bridging the Global North and South. Innovation fosters the knowledge economy, society and democracy. *The need to govern innovation may mean to design the governance of innovation in a way that it has the capability and capacity to govern a "self-different fractal innovation system".*

The "conceptual experiment" here was to focus on innovation and higher education systems practically to introduce (1) the "entrepreneurial university" as the "thesis", (2) the "academic firm" as the "antithesis", (3) and "epistemic governance" as the synthesis, as this governance approach in the context of *wider governance communities* is inclined to endorse and encourage further evolution of knowledge and innovation (see again Figure K). *Dialectics of thinking and "self-different fractal systems" are at the crossroads of reflecting and engaging in acting for innovation governance.*

The ideas discussed here aim to shape a vision for creating a ***"Manifesto of Higher Education Governance".***[64]

To reiterate our research question is: *How can the concept of a "self-different fractal innovation system" be applied to describe the interlinking of "entrepreneurial university," "academic firm," and "epistemic governance" in higher education governance?*

As a provisional answer, our analysis concludes with a "Conceptual Glossary", to further discussions on key concepts related to "self-different" fractal innovation systems.

The **Conceptual Glossary:**

1. **"Self-Different Fractals", "Dialectics of Innovation":** Fractals are generally seen as representations of "self-similar structures". The decision was made to explicitly introduce the concept of "self-different fractals". Self-different fractals are presented as a philosophical concept, directly connecting and aligning with

[64] The "Manifesto for Higher Education Governance" in the introduction we twisted now into the "Manifesto of Higher Education Governance".

the principles of the "Philosophy of Dialectics". *The interaction between "thesis" and "antithesis" could be disruptive, but when effectively managed and governed, it should foster creativity, co-creation, innovation, co-development, and finally development. Self-different fractal systems, being co-modal or multimodal, have a greater innovation potential.* Their interaction across various modes can spark creativity, leading to new ways of creating, producing, and applying knowledge. *Governing innovation may require designing governance structures capable of managing "self-different fractal systems of innovation" effectively.* This conceptual experiment, being presented here, focuses on innovation and higher education systems by introducing the entrepreneurial university as the "thesis", the academic firm as the "antithesis", and epistemic governance as the "synthesis".

2. **Research, Interdisciplinarity and Transdisciplinarity, Scientific Research and Artistic Research:** *Research (R&D, research and experimental development) involves knowledge creation and knowledge production.* Various research modes co-exist and evolve together. New research forms explore new forms of knowledge production. "Basic research in the context of application" (Campbell and Carayannis, 2013, p. 34) highlights further research prospects. *Interdisciplinary research tackles complex issues, themes and problems that one discipline alone cannot solve.* In transdisciplinary research, interdisciplinary academic researchers collaborate with user communities to ensure research applicability. *Transdisciplinarity may be defined as "context for the application of interdisciplinarity".*[65] *Scientific research and artistic research can engage in modes and networks of interdisciplinarity and transdisciplinarity.*
3. **Innovation, Linear and Non-Linear Innovation, Multi-Level Innovation Systems, Quadruple and Quintuple Helix Innovation Systems:** *Innovation associates with knowledge application.* Innovation is a process, "where knowledge is used for the purpose of a new application and new knowledge is being created" (Campbell and Pantelić, 2020a, p. 3). Learning processes in connection with innovation imply that "an improvement, betterment, advancement, or a reform are being achieved" (Campbell and Pantelić, 2020a, p. 3). *Linear innovation implies a sequential "first-then" relationship of*

[65] "In short, *transdisciplinarity may be defined as the application of interdisciplinarity* (transdisciplinarity = application of interdisciplinarity?). The Mode 3 inclination for a basic research in the context of application highlights a possible route of further development for transdisciplinarity and interdisciplinarity" (Campbell and Carayannis, 2013, pp. 34-35).

the different research phases, such as basic research, applied research, and application or implementation. In a non-linear innovation arrangement, the different research phases (basic research and application) operate "in-parallel". Innovation systems aggregate across different levels. *"Multi-level innovation systems" scale from local, regional, national and supranational, international and global.*[66] Universities (higher education institutions) in the sciences and arts are key participants in innovation systems. *The theory and concepts of "Quadruple and Quintuple Helix innovation systems" integrate the* "Triple Helix" (Etzkowitz and Leydesdorff, 2000) of economy, state, and higher education *within the context of the knowledge democracy and natural environments* (Carayannis and Campbell, 2009 and 2010; Carayannis et al., 2012). This also emphasizes the interactions between Industry 5.0, Society 5.0, Democracy 5.0 (Campbell and Carayannis, 2023b) and "Ecology 5.0". Veld (2010) was one of the first scholars to explicitly propose the concept of "knowledge democracy".

4. **Entrepreneurial University, Mode 3 University, Third Mission:** *The "entrepreneurial university" (entrepreneurial higher education institution) is a key concept of Third Mission* (Compagnuccia, and Spigarelli, 2020). "Mode 1" involves traditional discipline-specific academic research focused on basic research without practical application. "Mode 2" focuses on interdisciplinary and transdisciplinary research that involves diverse organizations, social responsibility, and quality standards (Gibbons et al., 1994). *"Mode 3" higher education institutions* (Campbell and Carayannis, 2013; Carayannis and Campbell, 2009) *blend traditional academic research from Mode 1 with the applied, interdisciplinary and transdisciplinary focus of Mode 2 to foster innovation. The Mode 3 knowledge paradigms encourage universities to decide whether to focus on Mode 1, Mode 2, or a mix of both while finding innovative ways to integrate them.* It promotes creative combinations of these modes, motivating institutions to overcome limitations and discover new ways to enhance quality and drive progress. *The "Mode 3 University" (or Mode 3 Higher Education Institution) concept encompasses an epistemic and organizational strategy aimed at synthesizing and evolving Mode 1 and Mode 2 knowledge production (networked knowledge paradigms), with the goal of generating*

[66] On "Multi-Level Systems of Innovation" see Kaiser and Prange-Gstöhl (2013).

a valuable surplus in knowledge.[67] Kuhn (1962) discusses the "structure of scientific revolutions": The epistemic diversity of knowledge and innovation modes within the organizational context of the Mode 3 university exemplifies why reflections on underlying epistemic understanding are interconnected with organizational development and governance. Mode 2 represents a model for the entrepreneurial university. However, the entrepreneurial university is not restricted or limited to Mode 2. Both Mode 2 and Mode 3 *universities (higher education institutions) are equipped to evolve and develop as advanced "entrepreneurial" universities. The Mode 3 university has the advantage of functioning as an internal ecosystem or environment with diverse knowledge and innovation modes.* The entrepreneurial university is not limited to just a few disciplines like business administration, economics and technological studies. *The whole disciplinary spectrum (including the social sciences, arts and humanities) has the potential of contributing to an entrepreneurial university. "Being entrepreneurial" means being involved in "being innovative" (as a person, as an organization),* with all the associated opportunities.

5. **Academic Firm:** *The "Academic Firm"* (Campbell and Güttel, 2005; Campbell and Carayannis, 2016a and 2023a; Campbell et al., 2013) *creates an "internal academic environment" or "internal academic ecosystem" within a company. The purpose of this is to simulate and emulate an "academic world" for a company, so to encourage and to facilitate similar motivations (as in the context of higher education) for engaging knowledge workers in research and innovation, where activities are connected with creativity-initiated and innovation-driven knowledge creation, production and application. and which may be translated into commercially successful business opportunities.* This should stimulate formation processes of "Creative Knowledge Environments" (Hemlin et al., 2004) within companies. The academic firm is deeply rooted in knowledge, prioritizing its creation and application. *It represents a distinct organization that is driven by the pursuit of knowledge production and application, particularly through research and innovation. The academic firm integrates academic principles to inspire its employees and foster strong bonds of trust and mutual support between the organization and its employees.* It sees the

[67] In an artistic installation a *"Picture of Fractals"* may be designed, where the different knowledge and innovation modes as an arrangement of "mosaic pictures". Such *"fractalized pictures"* can be drawn statically, or they are animated as "moving pictures".

diverse disciplinary backgrounds and diverse skills and competencies of its employees as advantageous, enhancing their ability to creatively contribute to knowledge production and application. The academic firm supports and promotes academic research, using academic publications as incentives to motivate employees, also to encourage employees to codify their "tacit knowledge" (Polanyi, 1966). *Partnerships with universities, higher education institutions and other academic research institutions are being activated so to access and participate in wider networks of knowledge production (research) and knowledge application (innovation).* Concerning working hours and schemes, the academic firm tries to be flexible, also with regard to "home-office" and "office-office", wherever feasible. This includes options for full-time, part-time, fluctuating schedules, as well as accommodating partial absences or leaves. *Working spaces may be physically close, or virtually remote (national, transnational, global). The academic firm often allows and occasionally encourages employees to hold dual (multiple) employment or "cross-employment" with academic institutions like universities or other organizations within or outside of higher education.*[68] The academic firm regards "continuing education", further education and lifelong learning as ways to support their knowledge workers and employees, and *furthermore is inclined to link internal promotion with participation in such (external) educational programs*. The academic firm can take or manifest in different organizational designs or settings, which include a whole firm, subunit, subdivision, or branch of a larger "commercial" firm focused on certain structures or processes, e.g. continuing education.

6. **The Co-Evolution of Entrepreneurial University and Academic Firm in Cross-Connecting and Inter-Linking Networks:** The unfolding of entrepreneurial university is being supported, if such a transition is being accompanied by an unfolding of academic firms, and when these organizations interact in cross-connecting networks. *Their good governance should yield innovation. Progress in IT and AI advances the capability and quality of networks to grow and to enhance the formation of innovative network structures and*

[68] This may also refer to "cross-retirement", for example, integrating or re-integrating the elderly knowledge workers into active work (Blasche and Campbell, 2013). The "unconditional basic income" represents an interesting concept (Prainsack, 2020). Possible cross-implications with cross-retirement require further thinking. All of this ought to be assessed in a wider framing of labor markets (Bast et al., 2019; Poier and Prettenthaler, 2007).

processes for "tomorrow" and beyond-tomorrow toward a new and "next" future.

7. **Governance, Epistemic Governance, New Public Management and Network Governance, Pluralism of Governance, Knowledge Democracy and Knowledge State:** *Higher education governance, both internal and external, addresses and supports the structures and processes within universities (or higher education institutions) for purposes of enhancing organizational development, also continuing development and ensured sustainable development.* Governance refers explicitly to the basic functions of universities : teaching, education and learning; research; Third Mission (entrepreneurial university, innovation, democracy education, and community engagement); possibly also a Fourth Mission ("higher education as co-creator of society and democracy"). New Public Management and Network Governance represent dominant approaches in the governance of higher education. Epistemic Governance is a further model of governance of and in higher education. *Epistemic Governance* (Campbell and Carayannis, 2013 and 2016b) *specifically refers to the underlying epistemic structure of academia, for example of "quality", and specifically designs and implements policy and strategy so to create an interlinkage between governance and organizational development. The governance of higher education takes place in the context of "knowledge democracies"* (Veld, 2010) *but also in context of the "Knowledge State"* (Campbell, 2006a, pp. 25-27). Furthermore, higher education is contextualized by "multi-level innovation systems" (local, regional, national, supranational, international and global). *National higher education systems are being embedded in international and global systems.* Governance increasingly takes network structures and network processes into account. Higher education governance addresses higher education institutions as concrete organizations, however, also sees higher education institutions as belonging to networks. *So. there is an overlap between institutional governance and network governance.* Internally, higher education institutions resemble, at least to a certain degree, "internal networks". Therefore, "external network governance" meets "internal network governance". *Organizational and national "boundaries" of universities are expressing fluid characteristics.* Diversity of knowledge and innovation, and a *pluralism of knowledge and pluralism of innovation* may be paraphrased as a "Democracy of Knowledge" (Carayannis and

Campbell, 2009, pp. 207-208).[69] *All together, this diversity of knowledge and innovation requires a "pluralism of governance"* (Campbell, 2024). *Pluralism of knowledge interlinks with pluralism of governance.*

8. **The Future of Governance of Higher Education:** *The future of higher education is being reflected in the future of its governance, of and in higher education.*[70] The future of governance is also the future of higher education.

Self-Different Fractals: Images by Jürgen Pirecki-Giefing.

[69] "This all may result in a 'democracy of knowledge', driven by a pluralism of knowledge and innovation and by a pluralism of paradigms for knowledge modes" (Carayannis and Campbell, 2009, p. 207).

[70] The "utopia" of future governance of higher education creates visions for higher education development. Refer to Schaper Rinkel (2020).

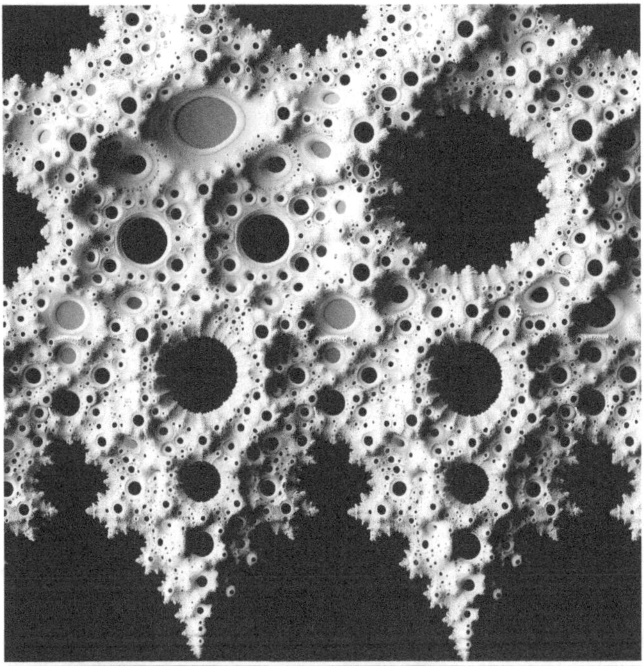

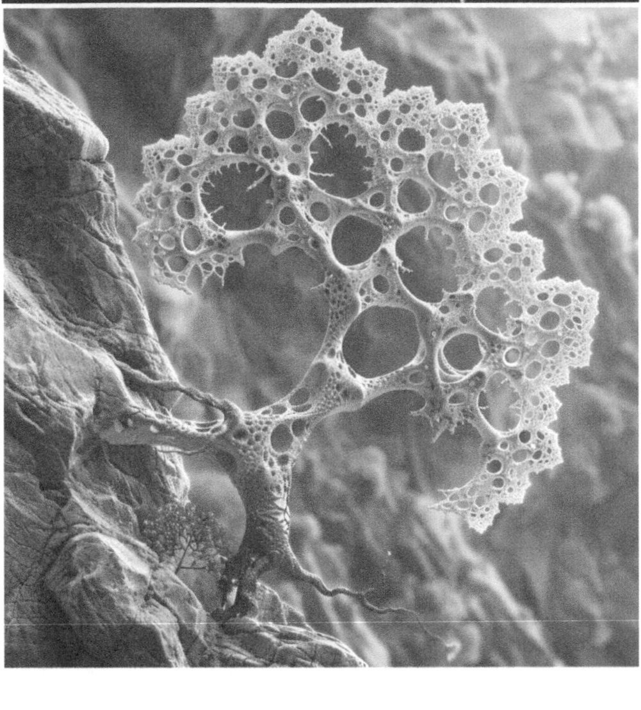

Fig. Cluster 2 Images of "Self-Different" Fractals
Author & Creator: Jürgen Pirecki-Giefing.

References

Arnold, Markus (2009). Interdisziplinarität: Theorie und Praxis eines Forschungskonzepts, 65–97, in: Markus Arnold (ed.): iff. Interdisziplinäre Wissenschaft im Wandel. Vienna: LIT.

Bast, Gerald / Elias G. Carayannis / David F.J. Campbell (eds.) (2019). The Future of Education and Labor. New York, NY: Springer (https://www.springer.com/gp/book/9783030260675).

Biegelbauer, Peter (ed.) (2010). Steuerung von Wissenschaft? Die Governance des österreichischen Innovationssystems. Innsbruck: Studienverlag.

Biegelbauer, Peter (2013). Wie lernt die Politik? Lernen aus Erfahrung in Politik und Verwaltung. Wiesbaden: Springer VS (https://doi.org/10.1007/978-3-531-19128-7).

Blasche, Gerhard W.E. / David F.J. Campbell (2013). Cross-Retirement (Cross-Employed Cross-Retired) and Innovation, 508–513, in: Elias G. Carayannis (Editor-in-Chief): Encyclopedia of Creativity, Invention, Innovation and Entrepreneurship. New York, NY: Springer (https://doi.org/10.1007/978-1-4614-3858-8_255).

Brandenburger, Adam M. / Barry J. Nalebuff (1997). Co-Opetition. New York, NY: Doubleday.

Brechelmacher, Angelika / Elke Park / Gülay Ates / David F.J. Campbell (2015). The Rocky Road to Tenure – Career Paths in Academia, 13-40, in: Tatiana Fumasoli / Gaële Goastellec / Barbara M. Kehm (eds.): Academic Work and Careers in Europe: Trends, Challenges, Perspectives. Cham, Switzerland: Springer (https://doi.org/10.1007/978-3-319-10720-2_2 and http://link.springer.com/chapter/10.1007/978-3-319-10720-2_2).

Burkert, Günther R. (2021). Die vernetzte Universität. Von der Kritik der Ökonomisierung zur Neuausrichtung auf die Gesellschaft. [The Networked University. From the Critique of Economization to a Reorientation Toward Society]. Vienna: Passagen.

Burkert, Günther R. / David F.J. Campbell / Thorsten D. Barth (2016). Die Vermessenheit der Vermessung: Überlegungen zur universitären Governance in den Geistes- und Sozialwissenschaften. [Measuring Measurement: Reflections on University Governance in the Humanities and Social Sciences]. *fteval Journal for Research and Technology Policy Evaluation* 42 (November 2016), 17–24 (http://www.fteval.at/upload/fteval_Journal_for_Research_and_Technology_Policy_Evaluation_42.pdf).

Burkert, Günther R. / David F.J. Campbell / Attila Pausits / Sina Westa (eds.) (2024). Zukunft. Hochschulen. Governance. Kremser Thesen. [Future. Higher Education Institutions. Governance. Kremser Theses]. Vienna: Passagen (https://doi.org/10.5771/9783709250822 and https://www.nomos-elibrary.de/10.5771/9783709250822.pdf?download_full_pdf=1).

Bush, Vannevar (1945). Science: The Endless Frontier. Washington, D.C.: United States Government Printing Office (http://www.nsf.gov/od/lpa/nsf50/vbush1945.htm#transmittal).

Cai, Yuzhuo / Annina Lattu (2021). Triple Helix or Quadruple Helix: Which Model of Innovation to Choose for Empirical Studies? *Minerva,* 60, 257–280 (https://doi.org/10.1007/s11024-021-09453-6 and https://link.springer.com/article/10.1007/s11024-021-09453-6).

Campbell, David F.J. (2000). Forschungspolitische Trends in wissenschaftsbasierten Gesellschaften. Strategiemuster für entwickelte Wirtschaftssysteme. *Wirtschaftspolitische Blätter* 47 (2), 130–143.

Campbell, David F.J. (2001). Politische Steuerung über öffentliche Förderung universitärer Forschung? Systemtheoretische Überlegungen zu Forschungs- und Technologiepolitik. Österreichische Zeitschrift für Politikwissenschaft 30, 425–438 (http://www.oezp.at/pdfs/2001-4-04.pdf).

Campbell, David F.J. (2003). The Evaluation of University Research in the United Kingdom and the Netherlands, Germany and Austria, 98–131, in: Philip Shapira / Stefan Kuhlmann (eds.): Learning from Science and Technology Policy Evaluation: Experiences from the United States and Europe. Camberley: Edward Elgar.

Campbell, David F.J. (2006a). Nationale Forschungssysteme im Vergleich. Strukturen, Herausforderungen und Entwicklungsoptionen. *Österreichische Zeitschrift für Politikwissenschaft* 35 (1), 25–44 (http://www.oezp.at/pdfs/2006-1-2-Campbell.pdf).

Campbell, David F.J. (2006b). The University/Business Research Networks in Science and Technology. Knowledge Production Trends in the United States, European Union and Japan, 67–100, in: Elias G. Carayannis / David F.J. Campbell (eds.): Knowledge Creation, Diffusion, and Use in Innovation Networks and Knowledge Clusters. A Comparative Systems Approach across the United States, Europe and Asia. Westport, Connecticut: Praeger.

Campbell, David F.J. (2011). Wissenschaftliche "Parallelkarrieren" als Chance. Wenn Wissenschaft immer öfter zur Halbtagsbeschäftigung wird, könnte eine Lösung im "Cross-Employment" liegen. Guest Commentary for DIE PRESSE (February 2, 2011) (http://diepresse.com/home/bildung/meinung/635781/Wissenschaftliche-Parallelkarrieren-als-Chance?direct=635777&_vl_backlink=/home/bildung/index.do&selChannel=5 Campbell, David F. J. (2013a). New University Governance: How the Academic Profession Perceives the Evaluation of Research and Teaching, 205–228, in: Ulrich Teichler / Ester Ava Höhle (eds.): The Work Situation of the Academic Profession in Europe: Findings of a Survey in Twelve Countries. Dordrecht: Springer (https://doi.org/10.1007/978-94-007-5977-0_10 and http://www.springer.com/education+%26+language/book/978-94-007-5976-3).

Campbell, David F.J. (2013b). Cross-Employment, 503-508, in: Elias G. Carayannis (Editor-in-Chief) / Igor N. Dubina, Norbert Seel, David F.J. Campbell, Dimitri Uzunidis (Associate Editors): Encyclopedia of Creativity, Invention, Innovation and Entrepreneurship. New York, NY: Springer (https://doi.org/10.1007/978-1-4614-3858-8_254).

Campbell, David F.J. (2017). Wie kann eine Organisation eine Qualitätskultur fördern? Erfahrungen, Lernprozesse und Zugänge an der Angewandten [How Can an Organization Foster a Quality Culture? Experiences, Learning Processes, and Approaches at the University of Applied Arts], 157–161, in: AQ Austria (ed.). Qualitätskultur. Ein Blick in die gelebte Praxis der Hochschulen. Beiträge zur 4. AQ Austria Jahrestagung 2016. [Quality Culture: A Look into the Lived Practices of Higher Education Institutions. Contributions to the 4th AQ Austria Annual Conference 2016]. Vienna: Facultas (https://www.pedocs.de/volltexte/2017/12911/pdf/AQ_Austria_2016_Qualitaetskultur_Tagungsband.pdf).

Campbell, David F.J. (2019). Global Quality of Democracy as Innovation Enabler. Measuring Democracy for Success. New York, NY: Palgrave Macmillan (https://link.springer.com/book/10.1007/978-3-319-72529-1).

Campbell, David F.J. (2024). Ein Pluralismus von Governance in der Governance ist wichtig, um Universitäten und die Hochschulsysteme in der Entwicklung "nach dem New Public Management" nachhaltig und innovativ zu unterstützen [A Pluralism of Governance in Governance Is Important To Support Universities and Higher Education Systems "After the New Public Management" in Sustainable and Innovative Ways], 29–39, in: Günther R. Burkert / David F. J. Campbell / Attila Pausits / Sina Westa (eds.) (2024). Zukunft. Hochschulen. Governance. Kremser Thesen. [Future. Higher Education Institutions. Governance. Kremser Theses]. Vienna: Passagen (https://www.nomos-elibrary.de/10.5771/9783709250822-29/ein-pluralismus-von-governance-in-der-governance-ist-wichtig-um-universitaeten-und-die-hochschulsysteme-in-der-entwicklung-nach-dem-new-public-management-nachhaltig-und-innovativ-zu-unterstuetzen?page=1).

Campbell, David F.J. / Wolfgang H. Güttel (2005). Knowledge Production of Firms: Research Networks and the "Scientification" of Business R&D. *International Journal of Technology Management* 31 (1/2), 152–175 (http://www.inderscience.com/search/index.php?action=record&rec_id=6629&prevQuery=&ps=10&m=or).

Campbell, David F.J. / Elias G. Carayannis (2011). Guest Editorial Preface. *International Journal of Social Ecology and Sustainable Development* 2 (1), i-vi (http://www.igi-global.com/Files/Ancillary/1947-8402_2_1_Preface.pdf).

Campbell, David F.J./Elias G. Carayannis (2012). Lineare und nicht-lineare knowledge production: innovative Herausforderungen für das Hochschulsystem. *Zeitschrift für Hochschulentwicklung* 7 (2), 64–72 (http://www.zfhe.at/index.php/zfhe/article/view/448).

Campbell, David F.J. / Elias G. Carayannis (2013). Epistemic Governance in Higher Education. Quality Enhancement of Universities for Development. SpringerBriefs in Business. New York, NY: Springer (http://www.springer.com/business+%26+management/organization/book/978-1-4614-4417-6).

Campbell, David F.J. / Elias G. Carayannis / Wolfgang H. Güttel (2013). Academic Firm, 17-23, in: Elias G. Carayannis (Editor-in-Chief) / Igor N. Dubina, Norbert Seel, David F.J. Campbell, Dimitri Uzunidis (Associate Editors): Encyclopedia of Creativity, Invention, Innovation and Entrepreneurship. New York, NY: Springer (http://link.springer.com/referenceworkentry/10.1007/978-1-4614-3858-8_252).

Campbell, David F.J. / Elias G. Carayannis / Scheherazade S. Rehman (2015). Quadruple Helix Structures of Quality of Democracy in Innovation Systems: the USA, OECD Countries, and EU Member Countries in Global Comparison. *Journal of the Knowledge Economy* 6 (3), 467-493 (https://doi.org/10.1007/s13132-015-0246-7 and (http://link.springer.com/article/10.1007/s13132-015-0246-7).

Campbell, David F.J. / Elias G. Carayannis (2016a). The Academic Firm: A New Design and Redesign Proposition for Entrepreneurship in Innovation-Driven Knowledge Economy. *Journal of Innovation and Entrepreneurship* 5:12 (pp. 1-10) (https://doi.org/10.1186/s13731-016-0040-1 and http://innovation-entrepreneurship.springeropen.com/articles/10.1186/s13731-016-0040-1).

Campbell, David F.J. / Elias G. Carayannis (2016b). Epistemic Governance and Epistemic Innovation Policy. *Technology, Innovation and Education* 2:2 (pp. 1-15) (https://doi.org/10.1186/s40660-016-0008-2 and http://technology-innovation-education.springeropen.com/articles/10.1186/s40660-016-0008-2).

Campbell, David F.J. / Elias G. Carayannis (2018). Overview of Cyber-Democracy, 323–326, in: Elias G. Carayannis / David F. J. Campbell / Marios Panagiotis Efthymiopoulos (eds.): Handbook of Cyber-Development, Cyber-Democracy, and Cyber-Defense. New York, NY: Springer (https://www.springer.com/de/book/9783319090689).

Campbell, David F.J. / Benjamin J.J. Bösch (2020). Future Innovations in Next Centuries, 1-6, in: Elias G. Carayannis (Editor-in-Chief): Encyclopedia of Creativity, Invention, Innovation and Entrepreneurship. New York, NY: Springer (Living Edition) (https://doi.org/10.1007/978-1-4614-6616-1_200088-1).

Campbell, David F.J. / Ivan Pantelić (2020a). Processes of Learning and Processes of Innovation, 1–6, in: Elias G. Carayannis (Editor-in-Chief): Encyclopedia of Creativity, Invention, Innovation and Entrepreneurship. New York, NY: Springer (Living Edition) (https://doi.org/10.1007/978-1-4614-6616-1_200098-1).

Campbell, David F.J. / Ivan Pantelić (2020b). Innovation of Joint Applications, 1-5, in: Elias G. Carayannis (Editor-in-Chief): Encyclopedia of Creativity, Invention, Innovation and Entrepreneurship. New York, NY: Springer (Living Edition) (https://doi.org/10.1007/978-1-4614-6616-1_200097-1).

Campbell. David F.J. / Kajetan Stransky-Can (2020). Von Governance zu Governance: Was kommt nach New Public Management? [From Governance to Governance: What Comes After New Public Management?] *Zeitschrift für Hochschulrecht* (zfhr) 19 (1), 54–62 (https://doi.org/10.33196/zfhr202002005401 and https://elibrary.verlagoesterreich.at/article/10.33196/zfhr202002005401).

Campbell, David F.J. / Günther R. Burkert / Thomas Lampoltshammer (2021). Vom Governance-zu-Governance-Lernen in Hochschulsystemen. Was kommt nach New Public Management? Jeder Hochschule ihre eigene Governance? Ein Opinion Piece. [From Governance-to-Governance Learning in Higher Education Systems. What Comes after New Public Management? For Every Higher Education Institution Her Own Governance? An Opinion Piece.], 203–213, in: Attila Pausits / Regina Aichinger / Martin Unger / Magdalena Fellner / Bianca Thaler (eds.): Rigour and Relevance. Hochschulforschung im Spannungsfeld zwischen Methodenstrenge und Praxisrelevanz. [Rigor and Relevance. Higher Education Research in the Field of Tension Between Strict Methodology and Practical Relevance.] Münster: Waxmann (https://www.waxmann.com/waxmann-buecher/?no_cache=1&tx_p2waxmann_pi2%5Bbuch%5D=BUC127851&tx_p2waxmann_pi2% 5Baction%5D=show&tx_p2waxmann_pi2%5Bcontroller%5D=Buch&cHash=33a5ad1129f105b369046e99b6054447 and https://www.waxmann.com/index.php?eID=download&buchnr=4459).

Campbell, David F.J. / Elias G. Carayannis (2023a). The Academic Firm Within a Cyber-D4 Environment, 300-310, in: Elias G. Carayannis / Evangelos Grigoroudis / David F.J. Campbell / Sokratis K. Katsikas (eds.): The Elgar Companion to Digital Transformation, Artificial Intelligence and Innovation in the Economy, Society and Democracy. Cheltenham: Edward Elgar (https://www.e-elgar.com/shop/gbp/the-elgar-companion-to-digital-transformation-artificial-intelligence-and-innovation-in-the-economy-society-and-democracy-9781839109355.html).

Campbell, David F.J. / Elias G. Carayannis (2023b). From Industry 4.0 to Democracy 5.0, 383–401, in: Elias G. Carayannis / Evangelos Grigoroudis (eds.): Handbook of Research on Artificial Intelligence, Innovation and Entrepreneurship. Cheltenham: Edward Elgar (https://www.elgaronline.com/edcollchap/book/9781839106750/book-part-9781839106750-38.xml and https://doi.org/10.4337/9781839106750.00038).

Campbell, David F.J. / Elias G. Carayannis (2023c). Digital Transformation of Democracy and Politics: Sustainable Development As A Crucial Dimension for Conceptualization and Measurement of Democracy and Quality of Democracy, 419–444, in: Elias G. Carayannis / Evangelos Grigoroudis (eds.): Handbook of Research on Artificial Intelligence, Innovation and Entrepreneurship. Cheltenham: Edward Elgar (https://www.elgaronline.com/edcollchap/book/9781839106750/book-part-9781839106750-40.xml and https://doi.org/10.4337/9781839106750.00040).

Campbell, David F.J. / Attila Pausits / Seamus Needham (2023d). The Role of Dual Education in Graduate Employability: The Comparison between Europe and Africa, 534–550, in: Tania Broadley / Yuzhuo Cai / Miriam Firth / Emma Hunt / John Neugebauer (eds.): The SAGE Handbook of Graduate Employability. SAGE Reference. London: SAGE Publications (https://doi.org/10.4135/9781529791082).

Campbell, George S. (2021). Wither the Westerly Wind. London: Austin Macauley.

Campbell, George S. / David F.J. Campbell (2011). The Semi-Aquatic Theory: Semi-Aquatic Evolutionary Phase and Environment, Language Development of Modern Humans. With a Short Epilog on Conceptualized Evolution, Social Ecology and the Quintuple Helix. *International Journal of Social Ecology and Sustainable Development* 2 (1), 15–30 (http://www.igi-global.com/bookstore/article.aspx?titleid=51634).

Carayannis, Elias G. / David F.J. Campbell (2006a). "Mode 3": Meaning and Implications from a Knowledge Systems Perspective, 1–25, in: Elias G. Carayannis / David F.J. Campbell (eds.): Knowledge Creation, Diffusion, and Use in Innovation Networks and Knowledge Clusters. A Comparative Systems Approach across the United States, Europe and Asia. Westport, Connecticut: Praeger.

Carayannis, Elias G. / David F.J. Campbell (2006b). Knowledge Creation, Diffusion, and Use in Innovation Networks and Knowledge Clusters. A Comparative Systems Approach across the United States, Europe and Asia. Westport, Connecticut: Praeger.

Carayannis, Elias G. / Christopher Ziemnowicz (2007). Rediscovering Schumpeter. Creative Destruction Evolving into "Mode 3". Houndmills: Palgrave Macmillan.

Carayannis, Elias G. / David F.J. Campbell (2009). "Mode 3" and "Quadruple Helix": Toward a 21st Century Fractal Innovation Ecosystem. *International Journal of Technology Management* 46 (3/4), 201-234 (https://doi.org/10.1504/IJTM.2009.023374 and https://www.inderscience.com/info/inarticle.php?artid=23374 and https://www.inderscience.com/offers.php?id=23374).

Carayannis, Elias G. / David F.J. Campbell (2010). Triple Helix, Quadruple Helix and Quintuple Helix and How Do Knowledge, Innovation and the Environment Relate To Each Other? A Proposed Framework for a Trans-disciplinary Analysis of Sustainable Development and Social Ecology. *International Journal of Social Ecology and Sustainable Development* 1 (1), 41–69 (https://doi.org/10.4018/jsesd.2010010105 and https://www.igi-global.com/article/triple-helix-quadruple-helix-quintuple/41959).

Carayannis, Elias G. / David F.J. Campbell (2011). Open Innovation Diplomacy and a 21st Century Fractal Research, Education and Innovation (FREIE) Ecosystem: Building on the Quadruple and Quintuple Helix Innovation Concepts and the "Mode 3" Knowledge Production System. *Journal of the Knowledge Economy* 2 (3), 327–372 (https://doi.org/10.1007/s13132-011-0058-3 and https://link.springer.com/article/10.1007/s13132-011-0058-3).

Carayannis, Elias G. / David F. J. Campbell (2012). Mode 3 Knowledge Production in Quadruple Helix Innovation Systems. 21st-Century Democracy, Innovation, and Entrepreneurship for Development. SpringerBriefs in Business. New York, NY: Springer (http://www.springer.com/business+%26+management/book/978-1-4614-2061-3).

Carayannis, Elias G. / Thorsten D. Barth / David F.J. Campbell (2012). The Quintuple Helix Innovation Model: Global Warming as a Challenge and Driver for Innovation. *Journal of Innovation and Entrepreneurship* 1 (1), 1–12. (https://doi.org/10.1186/2192-5372-1-2 and http://www.innovation-entrepreneurship.com/content/pdf/2192-5372-1-2.pdf).

Carayannis, Elias G. / Ali Pirzadeh / Denisa Popescu (2012b). Institutional Learning and Knowledge Transfer Across Epistemic Communities. New Tools of Global Governance. New York, NY: Springer (http://www.springer.com/economics/policy/book/978-1-4614-1550-3).

Carayannis, Elias G. / David F.J. Campbell (2014). Developed Democracies versus Emerging Autocracies: Arts, Democracy, and Innovation in Quadruple Helix Innovation Systems. *Journal of Innovation and Entrepreneurship* 3:12 (pp. 1–23) (https://doi.org/10.1186/s13731-014-0012-2 and http://www.innovation-entrepreneurship.com/content/3/1/12).

Carayannis, Elias G. / David F.J. Campbell (2015). Art and Artistic Research in Quadruple and Quintuple Helix Innovation Systems, 29-51, in: Gerald Bast / Elias G. Carayannis / David F.J. Campbell (eds.): Arts, Research, Innovation and Society. New York, NY: Springer (https://doi.org/10.1007/978-3-319-09909-5_3 and http://link.springer.com/chapter/10.1007/978-3-319-09909-5_3).

Carayannis, Elias G. / David F.J. Campbell (2017). Les systèmes d'innovation de la quadruple et de la quintuple hélice. [Quadruple and Quintuple Helix Innovation Systems.] *Innovations* 54 (3), 173–195 (https://www.cairn.info/revue-innovations-2017-3-p-173.htm).

Carayannis, Elias G. / David F.J. Campbell / Marios Panagiotis Efthymiopoulos (eds.) (2018). Handbook of Cyber-Development, Cyber-Democracy, and Cyber-Defense. New York, NY: Springer (https://www.springer.com/de/book/9783319090689).

Carayannis, Elias G. / David F.J. Campbell (2019). Smart Quintuple Helix Innovation Systems: How Social Ecology and Environmental Protection are Driving Innovation, Sustainable Development and Economic Growth. (SpringerBriefs in Business.). New York, NY: Springer (https://www.springer.com/de/book/9783030015169).

Carayannis, Elias G. / David F.J. Campbell (2020). The Knowledge and Innovation Principle of KITA PITA C2ITA, 1-5, in: Elias G. Carayannis (Editor-in-Chief): Encyclopedia of Creativity, Invention, Innovation and Entrepreneurship. New York, NY: Springer (Living Edition) (https://doi.org/10.1007/978-1-4614-6616-1_200100-1).

Carayannis, Elias G. / David F.J. Campbell / Evangelos Grigoroudis (2021). Democracy and the Environment: How Political Freedom Is Linked with

Environmental Sustainability. *Sustainability* 13 (10), 5522 (https://doi.org/10.3390/su13105522).

Carayannis, Elias G. / David F.J. Campbell (2021). Democracy of Climate and Climate for Democracy: the Evolution of Quadruple and Quintuple Helix Innovation Systems. *Journal of the Knowledge Economy* 12 (4), 2050–2082 (https://doi.org/10.1007/s13132-021-00778-x).

Carayannis, Elias G. / David F.J. Campbell (2022). Towards an Emerging Unified Theory of Helix Architectures (EUTOHA): Focus on the Quintuple Innovation Helix Framework as the Integrative Device. *Triple Helix* 2, 1–11 (https://doi.org/10.1163/21971927-bja10028 and https://brill.com/view/journals/thj/9/1/article-p65_6.xml).

Carayannis, Elias G. / David F.J. Campbell / Evangelos Grigoroudis (2022a). Helix Trilogy: the Triple, Quadruple, and Quintuple Innovation Helices from a Theory, Policy, and Practice Set of Perspectives. Journal of the Knowledge Economy 13 (3), 2272–2301 (https://doi.org/10.1007/s13132-021-00813-x and (https://link.springer.com/article/10.1007/s13132-021-00813-x).

Carayannis, Elias G. / Thorsten D. Barth / David F.J. Campbell (2022b). O MODELO DE INOVAÇÃO DA HÉLICE QUÍNTUPLA: O AQUECIMENTO GLOBAL COMO DESAFIO E MOTOR DA INOVAÇÃO. [The Quintuple Helix Innovation Model: Global Warming as a Challenge and Driver for Innovation.] *Revista Direito, Inovação e Regulações* / DOUTRINA INTERNACIONAL 1 (2), 89–111 (https://periodicos.univel.br/ojs/index.php/redir/article/view/178/133 and https://periodicos.univel.br/ojs/index.php/redir/article/view/178).

Carayannis, Elias G. / Joanna Morawska-Jancelewicz (2022). The Futures of Europe: Society 5.0 and Industry 5.0 as Driving Forces of Future Universities. *Journal of the Knowledge Economy* (online first) (https://doi.org/10.1007/s13132-021-00854-2).

Chux, Daniels / Christiane Gebhardt (2021). Special Issue: Higher Education, Science and Research Systems for Transformative Change in Africa – What Role for Innovation Policy? *Industry and Higher Education* 35 (5), 553-558 (https://doi.org/10.1177/09504222211029901).

Compagnuccia, Lorenzo / Francesca Spigarelli (2020). The Third Mission of the University: A Systematic Literature Review on Potentials and Constraints. *Technological Forecasting & Social Change* 161 (December 2020), 120284 (https://doi.org/10.1016/j.techfore.2020.120284).

Cunha, Miguel Pina e / João Vieira da Cunha / Stewart R. Clegg (2000). Management: Thesis, Antithesis, Synthesis. FEUNL Working Paper No. 395, Available at SSRN (https://ssrn.com/abstract=880465 and http://dx.doi.org/10.2139/ssrn.880465).

Danilda, Inger / Malin Lindberg / Britt-Marie Torstensson (2009). Women Resource Centres. A Quattro Helix Innovation System on the European Agenda. Paper (http://pure.ltu.se/portal/files/2806203/Danilda-Lindberg-Torstensson-paper.pdf).

Duberry, Jérôme (2022). Artificial Intelligence and Democracy. Risks and Promises of AI-Mediated Citizen–Government Relations. Cheltenham, UK: Edward

Elgar (file:///C:/Users/DCampbell/Downloads/9781788977319-9781788977319.pdf and https://www.elgaronline.com/monobook-oa/book/9781788977319/9781788977319.xmln).
Dubina, Igor N. / Elias G. Carayannis / David F. J. Campbell (2012). Creativity Economy and a Crisis of the Economy? Co-evolution of Knowledge, Innovation and Creativity, and of the Knowledge Economy and Knowledge Society. *Journal of the Knowledge Economy* 3 (1), 1–24 (http://www.springerlink.com/content/t5j8l12136h526h5/).
Ehlers, Ulf-Daniel (2020a). Future Skills. The Future of Learning and Higher Education. Karlsruhe: Self-Published & Books on Demand (https://nextskills.org/wp-content/uploads/2020/03/Future-Skills-The-Future-of-learning-and-higher-education.pdf).
Ehlers, Ulf-Daniel (2020b). Future Skills. Lernen der Zukunft – Hochschule der Zukunft. Wiesbaden: Springer VS (https://link.springer.com/book/10.1007%2F978-3-658-29297-3#toc).
Elfert, Maren / Christian Ydesen (2024). Global Governance and The Promissory Visions of Education: Challenges and Agendas. *Comparative Education* 60 (3), 361–376 (https://doi.org/10.1080/03050068.2024.2371701).
Elken, Mari (2024). Collaborative Design of Governance Instruments in Higher Education. *Studies in Higher Education* 49 (6), 1095–1106 (https://doi.org/10.1080/03075079.2023.2258905).
Enders, Jürgen / Jon File (eds.) (2010). Governance Reform. Progress in Higher Education Reform across Europe. Brussels: European Commission (http://ec.europa.eu/education/news/news2259_en.htm).
Ernst, Stefanie (2008). Manual Lehrevaluation. Wiesbaden: VS Verlag für Sozialwissenschaften.
Esping-Anderesen, Gøsta (1990). The Three Worlds of Welfare Capitalism. Princeton, New Jersey: Princeton University Press.
Etzkowitz, Henry / Loet Leydesdorff (2000). The Dynamics of Innovation: from National Systems and "Mode 2" to a Triple Helix of University-Industry-Government Relations. *Research Policy* 29, 109–123.
European Commission (2001a). Etymology of the Word Governance in the Different Languages. Étymologie du terme "gouvernance". Brussels: European Commission (http://ec.europa.eu/governance/docs/doc5_fr.pdf) (retrieved on December 29, 2010).
European Commission (2001b). European Governance. A White Paper. Brussels: European Commission (http://eur-lex.europa.eu/LexUriServ/site/en/com/2001/com2001_0428en01.pdf) (retrieved on December 29, 2010).
European Commission (2024). Community Engagement in Higher Education. Brussels: European Commission (https://community-engagement.eu/).
Faulhammer, Friedrich / Stefan Sagl (2024). "Strategischer Rahmen" an Stelle von "Strategie": Universitäre Governance muss perspektivenreicher gedacht werden, um strategische Ziele mit universitärer Autonomie wirksam zu verbinden ['Strategic Framework' instead of 'Strategy': University Governance must be Conceived with Broader Perspectives to Effectively

Align Strategic Goals with University Autonomy], 49–56, in: Günther R. Burkert / David F.J. Campbell / Attila Pausits / Sina Westa (eds.) (2024). Zukunft. Hochschulen. Governance. Kremser Thesen. [Future. Higher Education Institutions. Governance. Kremser Theses].

Fellner, Magdalena (2025). Studierfähigkeit als soziales Konstrukt. Eine gesellschaftskritische Perspektive. [The Ability to Study as a Social Construct: A Socio-Critical Perspective.] Frankfurt am Main: Campus (https://www.campus.de/buecher-campus-verlag/wissenschaft/soziologie/studierfaehigkeit_als_soziales_konstrukt-18441.html).

Ferlie, Ewan / Lynn Ashburner / Louise Fitzgerald / Andrew Pettigrew (1996). The New Public Management in Action. Oxford: Oxford University Press.

Ferlie, Ewan / Christine Musselin / Gianluca Andresani (2008). The Steering of Higher Education Systems: A Public Management Perspective. *Higher Education* 56 (3), 325–348 (http://www.springerlink.com/content/n22v78885l377144/fulltext.pdf).

Ferlie, Ewan / Christine Musselin / Gianluca Andresani (2009). The Governance of Higher Education Systems: A Public Management Perspective, 1-20, in: Catherine Paradeise / Emanuela Reale / Ivar Bleiklie / Ewan Ferlie (eds.). University Governance. Western European Comparative Perspectives. Dordrecht: Springer.

Geppert, Corinna / Attila Pausits / David F.J. Campbell / Florian Reisky / Franziska Lessky (2024). Die Akademische Profession in Österreich. Eine Analyse der APIKS-Austria-Erhebung. [The Academic Profession in Austria. An Analysis of the APIKS Austria Survey]. Münster: Waxmann (https://www.waxmann.com/index.php?eID=download&buchnr=4609).

Geuna, Aldo / Ben R. Martin (2003). University Researech Evaluation and Funding: An International Comparison. *Minerva* 41, 277–304 (http://www.personalweb.unito.it/aldo.geuna/Publications/Geuna%20and%20Martin%20Univ%20research%20evaluation_2003.pdf).

Gibbons, Michael / Camille Limoges / Helga Nowotny / Simon Schwartzman / Peter Scott / Martin Trow (1994). The New Production of Knowledge. The Dynamics of Science and Research in Contemporary Societies. London: Sage.

Goodwin, Mark (2009). Which Networks Matter in Education Governance? A Reply to Ball's New Philanthropy, New Networks and New Governance in Education. *Political Studies* 57, 680–687.

Gottweis, Herbert (1998). Governing Molecules. The Discursive Politics of Genetic Engineering in Europe and the United States. Cambridge, Massachusetts: MIT Press.

Graf, Siegfried (1987). Statistically Self-Similar Fractals. *Probability Theory and Related Fields* 74, 357–392 (https://doi.org/10.1007/BF00699096).

Hall, Peter A. (1993). Policy Paradigms, Social Learning, and the State: The Case of Economic Policymaking in Britain. *Comparative Politics* 25 (3), 275–296 (https://www.jstor.org/stable/422246).

Hanschitz, Georg Christoph (2023). The Bright Future of Ecosystem Economies: Explainable and Reliable Artificial Intelligence Via Software-Hardware Interoperability, 276–299, in: Elias G. Carayannis / Evangelos Grigoroudis / David F.J. Campbell / Sokratis K. Katsikas (eds.): The Elgar Companion to Digital Transformation, Artificial Intelligence and Innovation in the Economy, Society and Democracy. Cheltenham: Edward Elgar (https://www.e-elgar.com/shop/gbp/the-elgar-companion-to-digital-transformation-artificial-intelligence-and-innovation-in-the-economy-society-and-democracy-9781839109355.html).

Hemlin, Sven / Carl Martin Allwood/ Ben R. Martin (2004). Creative Knowledge Environments. The Influences on Creativity in Research and Innovation. Cheltenham: Edward Elgar.

Hooghe, Liesbet / Gary Marks (2001). Multi-level governance and European integration. Lanham: Rowman & Littlefield Publishers.

Huisman, Jeroen (2009). Coming to Terms with Governance in Higher Education, 1-9, in: Jeroen Huisman (ed.): International Perspectives on the Governance of Higher Education. Alternative Frameworks for Coordination. New York: Routledge.

Hutchinson, John E. (1981). Fractals and Self Similarity. *Indiana University Mathematics Journal* 30 (5), 713–747 (https://www.jstor.org/stable/24893080).

Ibanez, Alberto / Ahmed AlRadaideh / Juan Antonio Jimber del Rio / Gyanendra Singh Sisodia (2024). Good Governance and Innovation: a Renewed Global Framework for National and Supranational Policy Advancement. *Journal of the Knowledge Economy* 15, 5794–5816 (https://doi.org/10.1007/s13132-023-01324-7).

Janger, Jürgen / David F.J. Campbell / Anna Strauss (2019). Attractiveness of Jobs in Academia: A Cross-Country Perspective. *Higher Education* 78 (6), 991–1010 (https://doi.org/10.1007/s10734-019-00383-7 and https://link.springer.com/article/10.1007%2Fs10734-019-00383-7).

Kaiser, Robert / Heiko Prange-Gstöhl (2013). Multi-Level Systems of Innovation, 1316-1321, in: Elias G. Carayannis (Editor-in-Chief): Encyclopedia of Creativity, Invention, Innovation and Entrepreneurship. New York, NY: Springer (https://doi.org/10.1007/978-1-4614-3858-8_311).

Kallo, Johanna / Jussi Välimaa (2024). Anticipatory Governance in Government: The Case of Finnish Higher Education. Higher Education (https://doi.org/10.1007/s10734-024-01211-3).

Kickert, Walter (1995). Steering at a Distance: A New Paradigm of Public Governance in Dutch Higher Education. *Governance* 8 (1), 135-157 (http://onlinelibrary.wiley.com/doi/10.1111/j.1468-0491.1995.tb00202.x/abstract).

Kirat, I., Kocyigit, I. (2013) On the Dimension of Self-Affine Fractals. In: Stavrinides, S., Banerjee, S., Caglar, S., Ozer, M. (eds) Chaos and Complex Systems. Berlin and Heidelberg: Springer (https://doi.org/10.1007/978-3-642-33914-1_19).

Kritzinger, Sylvia / Barbara Prainsack / Helga Pülzl (2006). System oder Netzwerk? Veränderungen forschungspolitischer Strategien in Österreich. *Österreichische Zeitschrift für Politikwissenschaft* 35 (1), 75-92 (http://www.oezp.at/pdfs/2006-1-5-Kritzinge.pdf).

Krücken, Georg (2003a). Mission Impossible? Institutional Barriers to the Diffusion of the "Third Academic Mission" at German Universities. *International Journal of Technology Management* 25 (1/2), 18-33.

Krücken, Georg (2003b). Learning the "New, New Thing": On the Role of Path Dependency in University Structures. *Higher Education* 46 (3), 315-339.

Krücken, Georg / Frank Meier / Andre Müller (2007). Information, Cooperation, and the Blurring of Boundaries – Technology Transfer in German and American Discourses. *Higher Education* 53 (6), 675-696.

Kuhlmann, Stefan (1998). Politikmoderation. Evaluationsverfahren in der Forschung- und Technologiepolitik. Baden-Baden: Nomos.

Kuhn, Thomas S. (1962). The Structure of Scientific Revolutions. Chicago: The University of Chicago Press.

Langlet, Arne / Alice B.M. Vadrot (2024). Negotiating Regime Complexity: Following a Regime Complex in the Making. *Review of International Studies* 50 (2), 231–251 (https://doi.org/10.1017/S0260210523000402).

Larédo, Philippe / Philippe Mustar (eds.) (2001). Research and Innovation Policies in the New Global Economy. An International Comparative Analysis. Cheltenham: Edward Elgar.

Leydesdorff, Loet (2012). The Triple Helix, Quadruple Helix, …, and an *N*-tuple of Helices: Explanatory Models for Analyzing the Knowledge-based Economy? *Journal of the Knowledge Economy* 3 (1), 25–35 (http://www.springerlink.com/content/x543613918677871/).

Leydesdorff, Loet / Helen Lawton Smith (2021). Triple, Quadruple, and Higher-Order Helices: Historical Phenomena and (Neo-) Evolutionary Models. SSRN (http://dx.doi.org/10.2139/ssrn.3817410 and https://ssrn.com/abstract=3817410).

Luhmann, Niklas (1988). Neuere Entwicklungen in der Systemtheorie. Merkur 42 (4), 292–300.

Lundvall, Bengt-Åke (ed.) (1992). National Systems of Innovation. Towards a Theory of Innovation and Interactive Learning. London: Pinter Publishers.

Martin, Ben R. (2003). The Changing Social Contract for Science and the Evolution of the University, 7–29, in: Aldo Geuna / Ammon J. Salter / W. Edward Steinmueller (eds.): Science and Innovation: Rethinking the Rationales for Funding and Governance. Camberley: Edward Elgar.

Mittelstrass, Jürgen (2010). The Future of the University. *European Review* 18 (Supplement No. 1), S181–S189 (http://journals.cambridge.org/action/displayFulltext?type=1&fid=7451428&jid=ERW&volumeId=18&issueId=S1&aid=7451420&bodyId =&membershipNumber=&societyETOCSession=).

Müller, Karl H. (2018). Second-Order Science and New Cybernetics, 625–655, in: Carayannis, Elias G. / David F.J. Campbell / Marios Panagiotis

Efthymiopoulos (eds.): Handbook of Cyber-Development, Cyber-Democracy, and Cyber-Defense. New York, NY: Springer (https://doi.org/10.1007/978-3-319-09069-6_15).

Narin, Francis / Kimberley S. Hamilton / Dominic Olivastro (1997). The Increasing Linkage between U.S. Technology and Public Science. *Research Policy* 26, 317–330.

National Science Board (2010). Science and Engineering Indicators 2010. Arlington, VA: National Science Foundation. (http://www.nsf.gov/statistics/seind10/pdfstart.htm)

Neave, Guy (1988). On the Cultivation of Quality, Efficiency and Enterprise: An Overview of Recent Trends in Higher Education in Western Europe, 1986-1988. *European Journal of Education* 23 (1/2), 7–23.

Neave, Guy (1998). The Evaluative State Reconsidered. *European Journal of Education* 33 (3), 265–284.

Neave, Guy (2007). From Guardian to Overseer: Trends in Institutional Autonomy, Governance and Leadership. Paper presented at the CNE Conference on Institutional Autonomy, Governance and Leaderhip. Lisbon: CNE Conference.

Neave, Guy / Frans A. van Vught (eds.) (1991). Prometheus Bound: The Changing Relationship between Government and Higher Education in Western Europe. Oxford: Pergamon.

Nowotny, Helga / Peter Scott / Michael Gibbons (2001). Re-thinking science. Knowledge and the public in an age of uncertainty. Cambridge: Polity Press.

Nowotny, Helga / Peter Scott / Michael Gibbons (2003). Mode 2 Revisited: The New Production of Knowledge. *Minerva* 41, 179–194.

Nowotny, Helga / Peter Scott / Michael Gibbons (2006). Re-Thinking Science: Mode 2 in Societal Context, 39–51, in: Elias G. Carayannis / David F. J. Campbell (eds.): Knowledge Creation, Diffusion, and Use in Innovation Networks and Knowledge Clusters. A Comparative Systems Approach across the United States, Europe and Asia. Westport, Connecticut: Praeger.

OECD (1989). The Measurement of Scientific and Technical Activities. R&D Statistics and Output Measurement in the Higher Education Sector. Paris: OECD.

OECD (1994). Frascati Manual 1993. The Measurement of Scientific and Technological Activities. Proposed Standard Practice for Surveys of Research and Experimental Development. Paris: OECD.

OECD (2008). Tertiary Education for the Knowledge Society. Special Features: Governance, Funding, Quality. (Volume 1). Paris: OECD.

OECD (2024). Education at a Glance 2024. OECD Indicators. Paris: OECD (https://www.oecd.org/en/publications/education-at-a-glance-2024_c00cad36-en.html) .

Ouragini, Imen / Myriam Labidi / Anissa Ben Hassine Louzir (2024). The Role of Entrepreneurial Universities in Promoting Entrepreneurship Education: A Comparative Study Between Public and Private Technology Institutes.

Journal of the Knowledge Economy 15 (3), 10985–11006 (https://doi.org/10.1007/s13132-023-01528-x).

Page, Edward C. / Vincent Wright (eds.) (2007). From the Active to the Enabling State. The Changing Role of Top Officials in European Nations. London: Palgrave Macmillan.

Park, Elke (2013). From Academic Self-Governance to Executive University Management: Institutional Governance in the Eyes of Academics in Europe, 183-203, in: Ulrich Teichler / Ester Ava Höhle (eds.): The Work Situation of the Academic Profession in Europe: Findings of a Survey in Twelve Countries. Dordrecht: Springer (https://doi.org/10.1007/978-94-007-5977-0_9 and http://www.springer.com/education+%26+language/book/978-94-007-5976-3).

Pasternack, Peer (2008). Die Einheit von Forschung und Lehre. *DUZ Magazin* (2), 20–21. (http://www.peer-pasternack.de/texte/Einheit_von_Forschung_und_Lehre.pdf)

Pausits, Attila (2015). The Knowledge Society and Diversification of Higher Education: From the Social Contract to the Mission of Universities, 267–284, in: Adrian Curaj / Liviu Matei /Remus Pricopie /Jamil Salmi / Peter Scott (eds.): The European Higher Education Area. Between Critical Reflections and Future Policies. Cham: Springer (https://link.springer.com/content/pdf/10.1007%2F978-3-319-20877-0_18.pdf).

Pausits, Attila / Stefan Oppl / Sandra Schön / Magdalena Fellner / David F. J. Campbell et al (2023). Emergency and Innovation: The Impact of State-of-Emergency on Innovative Educational Practices During The COVID-19 Pandemic, 348–375, in: Elias G. Carayannis / Evangelos Grigoroudis / David F. J. Campbell / Sokratis K. Katsikas (eds.): The Elgar Companion to Digital Transformation, Artificial Intelligence and Innovation in the Economy, Society and Democracy. Cheltenham: Edward Elgar (https://www.e-elgar.com/shop/gbp/the-elgar-companion-to-digital-transformation-artificial-intelligence-and-innovation-in-the-economy-society-and-democracy-9781839109355.html).

Pechar, Hans / Lesley Andres (2011). Higher-Education Policies and Welfare Regimes: International Comparative Perspectives. *Higher Education Policy* 24 (March), 25–52 (http://www.palgrave-journals.com/hep/journal/v24/n1/full/hep201024a.html).

Peschke, Lutz (2021). The Prevention Paradox of the COVID-19 Crisis in Germany. Science Communication in Times of Uncertainties, 85–109, in: Sefer Darıcı / Ayşe Meriç Yazıcı, Ayşe (eds.): Coronalogy: Multidisciplinary Academic Analysis in Perspective of COVID-19. Warsaw: Sciendo. (https://doi.org/10.2478/9788366675162-005).

Peschke, Lutz / Sotorios Gyftopoulos, / Ayhan Kapusuzoğlu / Frans Folkvord et al (2023). Practices of Knowledge Exchange in the Context of the COVID-19 Pandemic. *Journal of the Knowledge Economy (15),* 12881–12918 (https://doi.org/10.1007/s13132-023-01537-w).

Pfeffer, Thomas (2012). Virtualization of Universities. Digital Media and the Organization of Higher Education Institutions. New York, NY: Springer (http://www.springer.com/business+%26+management/media+management/book/978-1-4614-2064-4).

Poier, Klaus / Franz Prettenthaler (eds.) (2007). Gerechte Arbeitswelt. Gobalisierung, Flexibilisierung, Armutsbekämpfung? Graz: Leykam.

Polanyi, Michael (1966). The Tacit Dimension. Chicago: University of Chicago Press (https://press.uchicago.edu/ucp/books/book/chicago/T/bo6035368.html).

Prainsack, Barbara (2020). Vom Wert des Menschen. Warum wir ein bedingungsloses Grundeinkommen brauchen. [On the Value of the Human Person. Why We Need an Unconditional Basic Income.] Vienna: Brandstätter Verlag (https://www.brandstaetterverlag.com/buch/vom-wert-des-menschen/).

Raab, Charles D. (1994). Theorizing the Governance of Education. *British Journal of Educational Studies* XXXXII (1), 6–22.

Rodrigues, Maria João (ed.) (2002). The New Knowledge Economy in Europe. A Strategy for International Competitiveness and Social Cohesion. Cheltenham: Edward Elgar.

Rothstein, Bo (2011). Quality of Government and Epistemic Democracy. (Paper to be presented at the Workshop *Epistemic Democracy in Practice* to be held at Yale University, New Haven, USA, Oct 20–22, 2011.) Gothenburg: University of Gothenburg (http://www.uio.no/english/research/interfaculty-research-areas/democracy/news-and-events/events/seminars/2011/papers-yale-2011/Yale-Rothstein.pdf).

Rhodes, Rod A. W. (1996). The New Governance: Governing without Government. *Political Studies* XLIV, 652–667 (http://law.hku.hk/gl/rhodes.pdf).

Rhodes, Rod A. W. (2008). Understanding Governance. Policy Networks, Governance, Reflexivity and Accountability. Maidenhead: Open University Press.

Rhodes, Rod A.W. / John Wanna / Patrick M. Weller (2009). Comparing Westminster. Oxford: Oxford University Press.

Schaper Rinkel, Petra (2020). Fünf Prinzipien für die Utopien von Morgen. [Five Principles for the Utopias of Tomorrow.] Vienna: Picus Verlag (https://www.picus.at/produkt/fuenf-prinzipien-fuer-die-utopien-von-morgen/).

Schibany, Andreas / Helmut Gassler (2010). Nutzen und Effekte der Grundlagenforschung. Vienna: Joanneum Research (http://www.bmwf.gv.at/fileadmin/user_upload/forschung/publikationen/Nutzen_und_Effekte_der_Grundlagenforschung.pdf).

Schwartzberg, Melissa (2015). Epistemic Democracy and Its Challenges. *Annual Review of Political Science* 18, 187–203 (https://doi.org/10.1146/annurev-polisci-110113-121908).

Scott, Peter (2009). Markets and New Modes of Knowledge Production, 58-77, in: Jürgen Enders / Egbert de Weert (eds.): The Changing Face of Academic Life. Analytical and Comparative Perspectives. London: Palgrave Macmillan.

Smith, Cameron / Michael Holden / Eustacia Yu / Patrick Hanlon (2021). "So What Do You Do?": Third Space Professionals Navigating a Canadian University Context. *Journal of Higher Education Policy and Management* 43 (5), 505–519 (https://doi.org/10.1080/1360080X.2021.1884513).

Sodaro, Michael J. (2004). Comparative Politics. A Global Introduction. Boston: McGraw Hill.

Sørensen, Mads P. / Lars Geschwind / Jouni Kekäle / Rómulo Pinheiro (eds.) (2019). The Responsible University. Exploring the Nordic Context and Beyond. Cham, Switzerland: Palgrave Macmillan (https://doi.org/10.1007/978-3-030-25646-3).

Stachowiak-Kudła, Monika / Sina Westa / Catarina Santos Botelho / Ildikó Bartha (2023). Academic Freedom as a Defensive Right. *Hague Journal on the Rule of Law,* 15, 161–190 (https://doi.org/10.1007/s40803-022-00188-4).

Tassey, Gregory (2000). Standardization in Technology-Based Markets. *Research Policy,* 29 (4-5), 587–602 (https://doi.org/10.1016/S0048-7333(99)00091-8).

Tassey, Gregory (2013). Technology Life Cycles, 1797–1807, in: Elias G. Carayannis (ed.): Encyclopedia of Creativity, Invention, Innovation and Entrepreneurship. New York, NY: Springer (https://doi.org/10.1007/978-1-4614-3858-8_460).

Tassey, Gregory (2018). Analysis of International Technology Development and Commercialization Policy Instruments and Infrastructures. University of Washington: Economic Policy Research Center (https://www.nist.gov/system/files/documents/2021/01/15/FOREIGN%20TBED%20Programs__Policy%20Targets%20and%20Impacts.pdf).

Teichler, Ulrich (2006). Was ist Qualität?, 168–184, in: Véronique Chalvet / Waldemar Dreger (eds.): Von der Qualitätssicherung der Lehre zur Qualitätsentwicklung als Prinzip der Hochschulsteuerung. Bonn: Hochschulkonferenz (Beiträge zur Hochschulpolitik).

Treib, Oliver / Holger Bähr / Gerda Falkner (2007). Modes of Governance: Towards a Conceptual Clarification. *Journal of European Public Policy* 14 (1), 1–20.

Umpleby, Stuart A. (1990). The Science of Cybernetics and the Cybernetics of Science. *Cybernetics and Systems* 21 (1), 109–121 (ftp://ftp.vub.ac.be/pub/projects/Principia_Cybernetica/Papers_Umpleby/Science-Cybernetics.txt) (retrieved December 29, 2010.

Vadrot, Alice B.M. (2008). "Der Wille zur Wahrheit als der Wille zur Macht". Die theoretischen Implikationen der Schnittstelle zwischen Politik und Wissenschaft in der Biodiversitätspolitik. (Magisterarbeit.) Vienna: University of Vienna.

Vadrot, Alice B.M. (2011). Reflections on Mode 3, the Co-evolution of Knowledge and Innovation Systems and How it Relates to Sustainable Development. Conceptual Framework for "Epistemic Governance". International Journal of Social Ecology and Sustainable Development 2 (1), 44–52 (http://www.igi-global.com/bookstore/article.aspx?titleid=51636).

Vadrot, Alice B.M. (2014). The Politics of Knowledge and Global Biodiversity. London: Routledge (https://www.routledge.com/The-Politics-of-Knowledge-and-Global-Biodiversity/Vadrot/p/book/9781138209879?srsltid=AfmBOopVq0pFOHxcLxBYmxvaRKXk_FYoya0nOfIRcYB_5ooNG6vbXZ3h).

Veld, Roeland J. in´t (2010). Knowledge Democracy. Consequences for Science, Politics, and Media. Heidelberg: Springer (https://link.springer.com/book/10.1007/978-3-642-11381-9).

Veles, Natalie / Margaret-Anne Carter / Helen Boon (2018). Complex Collaboration Champions: University Third Space Professionals Working Together Across Borders. *Perspectives: Policy and Practice in Higher Education* 23 (2-3), 75–85 (https://doi.org/10.1080/13603108.2018.1428694).

Wagner, Caroline S. / David Roessner / Kamau Bobba / Julie Thompson Klein / Kevin W. Boyack et al (2011). Approaches to Understanding and Measuring Interdisciplinary Scientific Research (IDR): A Review of the Literature. *Journal of Informetrics* 165, 14–26 (https://doi.org/10.1016/j.joi.2010.06.004 and https://www.researchgate.net/publication/229352533_Approaches_to_Understanding_and_Measuring_Interdisciplinary_Scientific_Research_IDR_A_Review_of_the_Literature).

Weber, Viktoria (2024). Begrüßung zu den "Kremser Hochschulgespräche" zu Governance und Transformation. [Welcome Address for the "Higher Education Talks in Krems" on Governance and Transformation]. Press Release by the University for Continuing Education Krems on September 12, 2024. Krems, Austria: University for Continuing Education Krems (https://www.donau-uni.ac.at/de/aktuelles/news/2024/kremser-hochschulgespraeche-zu-governance-und-transformation.html).

Welp-Park, Elke (2023). Governance of Academic Careers in Austria. A History of Career Paths, Modes of Employment and Appointment Procedures at Austrian Universities from 1365 to 1920. Dissertation. Klagenfurt: University of Klagenfurt.

Westa, Sina (2024). Die KI transformiert die Wissenschaft und ihre Governance und wird diese in Zukunft revolutionieren, wenn nicht sogar evolvieren [AI is Transforming Science and its Governance and will Revolutionize, if not Fundamentally Evolve, them in the Future], 85–92, in: Günther R. Burkert / David F.J. Campbell / Attila Pausits / Sina Westa (eds.) (2024). Zukunft. Hochschulen. Governance. Kremser Thesen. [Future. Higher Education Institutions. Governance. Kremser Theses]. Vienna: Passagen (https://www.nomos-elibrary.de/10.5771/9783709250822-29/ein-pluralismus-von-governance-in-der-governance-ist-wichtig-um-universitaeten-und-die-hochschulsysteme-in-der-entwicklung-nach-dem-new-public-management-nachhaltig-und-innovativ-zu-unterstuetzen?page=1).

Wiener, Norbert (1948). Cybernetics or Control and Communication in the Animal and the Machine. New York: John Wiley & Sons Inc.

Willke, Helmut (1983). Entzauberung des Staates. Überlegungen zu einer sozietalen Steuerungstheorie. Königstein/Ts.: Athenäum (http://www.uni-bielefeld.de/soz/globalgov/Lit/Willke_Entzauberung.pdf).

Willke, Helmut (1989). Systemtheorie entwickelter Gesellschaften. Dynamik und Riskanz moderner gesellschaftlicher Selbstorganisation. Weinheim: Juventa.

Willke, Helmut (1998). Systemtheorie III: Steuerungstheorie. Stuttgart: Lucius & Lucius.

Zenk, Lukas / Attila Pausits / Barbara Brenner / David F.J. Campbell / Doris A. Behrens et al (2024). Meta-Competences in Complex Environments: An Interdisciplinary Perspective. Thinking Skills and Creativity 53 (September 2024), 101515 (1–17) (https://doi.org/10.1016/j.tsc.2024.101515).

Ziegele, Frank (2024). Hochschulbildung wird zum lebenslangen Prozess. Aus zertifizierten Bausteinen lebenslanger akademischer und beruflich-praktischer Bildung bildet sich ein individuelles Kompetenzportfolio heraus [Higher Education is Becoming a Lifelong Process. An Individual Competency Portfolio is Formed from Certified Modules of Lifelong Academic and Practical Vocational Education], 57-62, in: Günther R. Burkert / David F. J. Campbell / Attila Pausits / Sina Westa (eds.) (2024). Zukunft. Hochschulen. Governance. Kremser Thesen. [Future. Higher Education Institutions. Governance. Kremser Theses]. Vienna: Passagen (https://www.nomos-elibrary.de/10.5771/9783709250822-29/ein-pluralismus-von-governance-in-der-governance-ist-wichtig-um-universitaeten-und-die-hochschulsysteme-in-der-entwicklung-nach-dem-new-public-management-nachhaltig-und-innovativ-zu-unterstuetzen?page=1).

Index

A

Academic Firm 1, 2, 4, 11, 14, 37, 39, 41, 42, 45, 51-56, 61, 62, 64, 65
Artistic Research 16, 31, 33, 38, 46, 57, 62

D

Dialectics of Innovation 1, 2, 4, 61

E

Entrepreneurial University 1, 4, 13-15, 39, 41, 45, 61-66
Epistemic Governance 1, 2, 4, 26, 28-33, 36, 44, 46, 48, 49, 61-62, 66

F

Future of Governance 18, 67

G

Governance 1, 2, 4, 8-12, 15, 17-25, 29, 30, 35, 46, 48, 50, 60-62, 64, 65, 67

I

Innovation 1-4, 11-13, 18, 20, 22-24, 30, 37-40, 42, 44, 45, 47, 49, 52, 54, 56, 57, 60, 61, 64, 66
Interdisciplinarity 14-16, 46, 50, 62

K

Knowledge Democracy 19, 24, 32, 40, 51, 56, 63, 66
Knowledge State 4, 19, 20, 24, 51, 66

L

Linear and Non-Linear Innovation 39, 40, 44, 62

M

Mode 3 University 4, 46, 48, 63, 64
Multi-Level Innovation Systems 46, 62, 63

N

New Public Management 2, 10, 27, 28, 66
Network Governance 2, 10, 27, 28, 50, 66

P

Pluralism of Governance 2, 46, 48, 66, 67

Q

Quadruple and Quintuple Helix Innovation Systems 11, 51, 57, 62, 63

R

Research 1, 2, 11, 14, 16, 18, 20, 22, 23, 25, 27, 30, 32, 34, 37-43, 46-48, 51, 52, 54, 57, 59, 62-65

S

Self-Different Fractals 1, 3, 58, 59-61, 67, 70

T

Third Mission 2, 12-15, 23, 49, 50, 63, 66
Transdisciplinarity 15, 16, 42, 46, 50, 62